STUDY GUIDE WITH SELECTED SOLUTIONS
TO ACCOMPANY

WORLD OF CHEMISTRY
Essentials

SECOND EDITION

Walt Volland

BELLEVUE COMMUNITY COLLEGE,
WASHINGTON

SAUNDERS GOLDEN SUNBURST SERIES

Saunders College Publishing
Harcourt Brace College Publishers

Fort Worth Philadelphia San Diego New York Orlando Austin
San Antonio Toronto Montreal London Sydney Tokyo

Copyright © 1999 Saunders College Publishing

All rights reserved. No part of this publication may be reproduced or transmitted in any form or by any means , electronic or mechanical, including photocopy, recording, or any information storage and retrieval system, without permission in writing from the publisher.

Requests for permission to make copies of any part of the work should me mailed to: Permissions Department, Harcourt Brace & Company, 6277 Sea Harbor Drive, Orlando, Florida 32887-6777.

Portions of this work were published in previous editions. (Permissions acknowledgments and other credits to appear here.)

Printed in the United States of America

ISBN 0-03-022392-X

901 202 765432

Contents

	Internet Basics	iii
1	Living in a World of Chemistry	1
2	The Chemical View of Matter	13
3	Atoms and the Periodic Table	31
4	Nuclear Changes	49
5	Chemical Bonding and States of Matter	63
6	Chemical Reactivity: Chemicals in Action	85
7	Acid - Base Reactions	105
8	Oxidation - Reduction Reactions	123
9	Energy and Hydrocarbons	133
10	Organic Chemicals and Polymers	145
11	The Chemistry of Life	159
12	Nutrition: The Basis of Healthy Living	179
13	Chemistry and Medicine	193
14	The Chemistry of Useful Materials	211
15	Water: Plenty of It, But of What Quality?	223
16	Air: The Precious Canopy	239
17	Feeding the World	257
	Speaking of Chemistry Solutions	273

Preface

This Study Guide is a little different from most. It is designed to be more than just a help to you in your present chemistry class. I intend that it bring some fun into the study of chemistry, to make chemistry enjoyable and less threatening or boring. Because I hope the activities will heighten your curiosity about the world, I have included a variety so that you can see the many ways that all of us are connected to chemistry. The Bridging the Gap activities, from the simplest, kitchen counter experiments to the more elegant Internet adventures, all lead you to a wider and more understandable world. I really hope that you take something of lasting value from this book. It might be the systematic methods that are part of problem solving; it might be an increased awareness of your surroundings; or it might be a continuing addiction to Internet surfing. These all can be useful for the rest of your life. Good luck in all you do.

Why things are way they are.
The summaries, objectives and key terms are aimed at helping you focus on the highlights contained in each section from the point of view of someone who knows, and actually enjoys chemistry. The additional readings are in all types of publications from *Mademoiselle* and *Glamour* to *The American Scientist*. When space allows, the answers to questions, include an explanation of the reasoning behind the answer. Similarly the solutions to problems are written to help you increase your analytical skills. The cross word puzzles are supposed to give you an entertaining way to build your chemistry vocabulary. The Bridging the Gap exercises are aimed at showing how your everyday activities are linked to chemistry and science. The Internet activities are a taste of the future. The movie "Waterworld" reminds us of the power of entertainment to shape our view of reality even though we know the movie is fiction.

Acknowledgments
I want to thank the authors of the World of Chemistry---Essentials, Mel Joesten and Jim Wood, and the Contributing Editor, Mary Castellion, for their confidence in me. I appreciate the help of Ed Dodd, Editorial Coordinator of Saunders College Publishing, for his patience and encouragement. This book follows in the foot steps of the World of Chemistry 2e. A special note of appreciation goes to, Beth Rosato, Senior Developmental Editor, who gave me the initial opportunity to work on the World of Chemistry 2e.

My deepest thanks go to my wife Gloria and my sons Kirk and Greg. Gloria's understanding, her knowledge of chemistry, help at proof reading, editing, and heroic support are what made this book possible. Kirk and Greg gave invaluable advice regarding the nonchemists view and the Internet exercises. They both helped keep the work in perspective.

Feedback
I am interested in receiving comments and suggestions. Please contact me by snail mail or email.

Dr. Walt Volland
Department of Chemistry
Bellevue Community College
3000 Landerholm Circle S.E.
Bellevue, WA 98007-6484

email: wvolland@bcc.ctc.edu
fax: 425-641-2230

WVV
Bellevue, Washington
June, 1998

© 1999 Harcourt Brace & Company. All rights reserved.

Internet Basics

How do you get on the Internet?
Some of you already have an Internet provider; like America Online, Prodigy, CompuServe. If you do not, you can use public Internet access facilities. Check your local public library and ask if library users have access to computer terminals with Internet service. Many colleges offer Internet access through college libraries, campus computer centers, departmental terminals or even have it wired to dormitory rooms. You should ask your instructor, college librarians, or someone in computer services about access on your campus. However you "get onto the Net", have confidence that you, too, can learn to use it. If you are mature enough to remember using the card files in a library, you can use the Net. If you can play a video game, you can use the Net. If you can record a TV show on your VCR, you can use the Net. Even if you never learned to type, you can use the Net.

There is a tremendous amount of trivial information on the Internet. It also provides a direct, quick connection to government agencies and other sources of hard factual information. The exercises in this book are designed to give you a reason to learn about the power of the Internet. The ease of contacting people and accessing information from all the corners of the world is breathtaking for most of us. You can read an article in London one minute and then see a satellite view of U.S. regional weather fronts from NOAA in the next.

How do you navigate on the Internet?
The Internet typically is accessed using a browser. Most browsers look a lot alike. One of the most common browsers is Netscape© which is a very good browser and is free to students, educators and schools. The *Show Location* window for Netscape© is shown below. The Uniform Resource Locators, URLs, given in the Internet assignments can be typed in the "Location:" window. The URL you see in this picture is for the United State Geological Survey. This is series of letters, colon, periods, and slash marks is the URL.

> http://h2o.usgs.gov/public/pubs/bat/fig6.gif

You can type in this URL, hit the return key, and you will soon see this USGS web page. It is critical that you type in everything correctly; missing a period, slash, or a letter will prevent your reaching the desired page. So will including "extra" letters, periods, or slashes.

The different command buttons can be explored if you want. For example the "Back" button lets you go back to a previously entered URL; it is much like turning back one page in a book. The "Net Search" button opens up various search engines that enable you to search the net like you would use your library computer search engine. The big difference is that on the Internet you are searching worldwide sources.

© 1999 Harcourt Brace & Company. All rights reserved.

What is on an Internet page besides information?

The Internet pages typically have interactive or "hot" buttons. The most common interactive spots are underlined text items. Sometimes there are images that are "hot" buttons. Often they are identified with "Click here" messages. Usually, when you move your cursor over a "hot" spot on a web page, the cursor shows the outline of a hand. You can't go wrong by clicking on an item to find out if it is active or "hot". ONE CAUTION: there is a time delay between the time you click and the time when the system responds and connects you. Many browsers tell you what is happening to your request in a small window at the bottom of the page. Netscape© shows an image that looks like comets and shooting stars in motion behind the large "N" in its logo when a request is being processed. YOU WILL NEED SOME PATIENCE HERE. Computers have gotten faster, but people are asking the systems to do more, and millions people are using the Internet. You probably already know you can't rush the machines! If you change your mind, you can always click on the "Stop" button.

This is a chance to surf the net. Go ahead. Get your feet wet and have fun!

Dr. V.

© 1999 Harcourt Brace & Company. All rights reserved.

1 Living in a World of Chemistry

1.1 The World of Chemistry

This section points out that all too often the word "chemical" is associated with hazardous or toxic substances. A point is made that everything, whether natural or man made, contains some chemical because all matter is some kind of chemical. The future of humanity and of the planet depend on the way in which chemicals are used. A scientific fact is defined as an experimental result that is repeatedly observed under the same conditions. The scientific method and the chemist's way of viewing the world are described.

Objectives

After studying this section a student should be able to:
- give examples of scientific facts
- tell why models and theories are developed
- tell what happens when experimental results disagree with results predicted by theory
- explain how observations and facts relate to models and theories

Key terms

- chemical
- scientific method
- models
- all natural products
- experiment
- theories
- scientific fact
- predictions

1.2 DNA Fingerprinting, Biochemistry and Science

This section explains how the field of DNA testing, often front page news in sensational criminal cases, is dependent on biochemistry. The historical and practical links between the physical sciences and biological sciences are described. A graphic shows the subdivisions of the natural sciences. Chemistry is identified as one of the four physical sciences: astronomy, physics, geology, and chemistry. The branches of chemistry are listed as analytical, physical, organic, and inorganic. The overlapping roles of physical and biological science in the study of DNA are highlighted.

Objectives

After studying this section a student should be able to:
- name the two branches of natural science
- give the definition for chemistry
- name four divisions of chemistry and tell what is studied in each
- name two disciplines of chemistry that deal with DNA testing

Key terms

- natural science
- chemistry
- organic chemistry
- botany
- synthesize new matter
- physical science
- biochemistry
- inorganic chemistry
- genetics
- DNA fingerprint
- biological science
- analytical chemistry
- physical chemistry
- natural matter

© 1999 Harcourt Brace & Company. All rights reserved.

1.3 Air Conditioning, the Ozone Hole, and Technology

This section recounts the historical development of chlorofluorocarbons as refrigerant gases and as propellants. The benefits and unforeseen consequences of their use are described. The ozone depletion problem is summarized. Basic science is defined as the pursuit of knowledge with no short term practical objectives. Applied science is science with the short term goal of solving a specific problem. Technology is defined as the application of scientific knowledge in society to address economic, social, and industrial issues. A brief summary shows changes in the incubation period of an invention from its creation to its commercial use.

Objectives
After studying this section a student should be able to:
- describe the history of chlorofluorocarbons from development to common use
- give definitions for basic science, applied science, technology
- describe the link between CFCs and ozone depletion
- describe how the incubation period for innovations has changed since 1900

Key terms
stable	applied and basic science	CFCs and refrigerants
technology	ozone	incubation period

1.4 Automobile Tires, Hazardous Waste, and Risk

This section describes the case of a fire in a scrap tire dump and discusses the concept of risk. The risks associated with tire disposal are discussed. The concepts of risk assessment and management are explained. Estimates of increased risk of death after engaging in various activities are tabulated. Governmental actions to attempt to regulate risks are described.

Objectives
After studying this section a student should be able to:
- give the definitions for risk assessment and risk management
- give the definitions for risk-based laws, and technology-based laws

Key terms
risk assessment	balancing laws	EPA
technology-based law	risk management	

1.5 What is Your Attitude Toward Chemistry?

This section discusses the reasons to study chemistry. The idea that protection of health, safety and the environment depend on a knowledge of chemistry is introduced.

Objectives
After studying this section a student should be able to:
- explain why chemophobia may make matters worse
- tell how a knowledge of chemistry can make decisions about health and the environment more effective

Key terms and equations
chemophobia

© 1999 Harcourt Brace & Company. All rights reserved.

Additional readings

Ausubel, J. H. "Can Technology Spare the Earth?" American Scientist Mar-April 1996: p 166.

Beardsley, Tim. "Death by Analysis" Scientific American June 1995 : p 28.

Gibbs, W. Wayt , "Ounce of Prevention: Cleaner chemicals pay off, but industry is slow to invest" Scientific American Nov 1994: p 103.

Horgan, John. "Radon's Risks: Is the EPA exaggerating the dangers of this ubiquitous gas?" Scientific American Aug. 1994: p 14.

Lander, Eric S. "DNA Fingerprinting on Trial" Nature June 15, 1989: p 505.

Neufeld, Peter J. and Neville Colman. "When Science Takes the Witness Stand" Scientific American May 1990: p 46.

"Putting Sunscreens to the Test" Consumer Reports May 1995: p 334.

"Secondhand Smoke, Is It a Hazard?" Consumer Reports Jan. 1995: p 27.

Zorpette, Glenn. "Bracing for the Next Big One: Engineers grapple with retrofitting Japanese and U.S. buildings" Scientific American April 1995: p14.

Speaking of Chemistry

Name _____

Living in a World of Chemistry

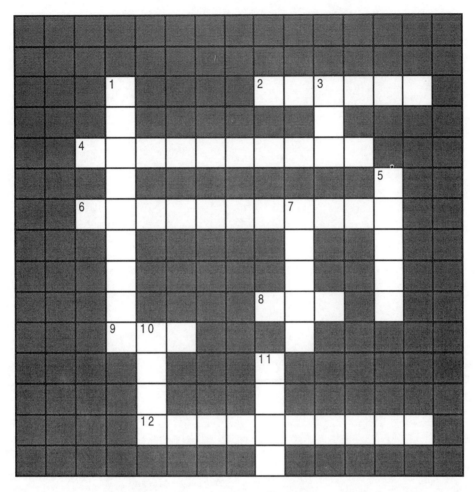

Across
2 Mixing ammonia and household _____ produces toxic chlorine gas
4 Chemistry that deals with the determination of structure and composition
6 An unreasonable fear of chemicals
8 Deoxyribonucleic acid
9 Chlorofluorocarbons used in refrigeration systems
12 The application of scientific knowledge to our society, economic system and industry.

Down
1 Materials made by human activity instead of natural processes.
3 Environmental Protection Agency
5 Type of science with no immediate goal
7 Gas in the stratosphere that is being depleted by interaction with CFCs
10 A repeatedly observed experimental result
11 Science that is not widely accepted nor developed with necessary care.

© 1999 Harcourt Brace & Company. All rights reserved.

Bridging the Gap I   Name_____
Risks: DNA Fingerprinting and Secondhand Smoke

This is a two part exercise. The first part addresses the issue of the admissibility of DNA fingerprinting as evidence in criminal cases and the second deals with secondhand smoke.

1.   Admissibility of DNA Fingerprinting evidence

Judicial decisions to admit laboratory test results as evidence in criminal cases are fraught with problems and risks for society. The traditional attitude in the criminal justice system in the United States accepts legal precedence as a basis for admissibility of evidence. Perhaps the most notorious example of the problem this poses is the so-called paraffin test which was used by crime laboratories throughout the U.S. to detect nitrite and nitrate residues, presumably from gunpowder, on suspects' hands to show that they had recently fired a gun. The test was first admitted as scientific evidence in a 1936 trial in Pennsylvania. Other states then adopted that decision without scrutinizing the supporting research. For 25 years numerous defendants were convicted with the help of this test. It was not until the mid-1960's that the test was subjected to careful scientific study; the study revealed a damning flaw in the test. It gave an unacceptably high number of false positives. Substances other than gunpowder that gave a positive reading included urine, tobacco, tobacco ash, fertilizer and colored fingernail polish. In this instance the legal process failed, allowing people to be convicted on scientific data that later proved to be worthless.

Pretend you are a judge. Develop two questions you would want answered by experts before you would decide whether or not DNA fingerprinting should be admissible as evidence in criminal cases. Include a reason for wanting each of these questions answered. Write these questions and reasons on this sheet.

© 1999 Harcourt Brace & Company. All rights reserved.

2. Secondhand Smoke

Decisions in daily life are made based on many factors. One such decision deals with the question of smoking. Personally each person can make their own choice, but the issue of secondhand smoke has created pressure for legislation to bar smoking in public places. Pretend you are a member of the state legislature. Your task is to develop two questions you would want answered by experts regarding the risks of secondhand smoke before you would decide on legislation. Include a reason for wanting each of these questions answered. Write these questions and reasons on this sheet.

Bridging the Gap II
Internet access to the Center for Disease Control, CDC and the Food and Drug Administration, FDA

There is a tremendous amount of trivial information on the Internet. It also provides a direct, quick connection to government agencies and other sources of hard factual information. This is an exercise in using the Internet to learn about the power of the Internet. Simultaneously, you will have an opportunity to learn what health agencies like the FDA and the CDC say about health risks and smoking.

The Internet pages typically have interactive or "hot" buttons. The most common interactive spots are underlined text items. Sometimes there are images that are buttons. Usually they are identified with "Click here" messages. You can't go wrong by clicking on an item to find out if it is active. ONE CAUTION, there is a time delay between the time you click and the time when the system responds and connects you. You need some patience here. Computers have gotten faster, but people are asking the computer to do more. You probably already know you can't rush the machines. This is a chance to surf the net. Go ahead; get your feet wet.

The CDC home page and the FDA
Public agencies like the Center for Disease Control, CDC, and the Food and Drug Administration, FDA, can be accessed through the Internet. The CDC page on tobacco risks can be reached by using the following uniform resource locator (URL). Your assignment is to open this page. The site will give you data to answer the questions on the report sheet.

http://www.cdc.gov/nccdphp/osh/issue.htm

The CDC site will have a heading like the one shown below.

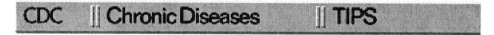

OVERVIEW

You are to find out what the CDC says about the number of deaths annually in the United States that are said to be caused by smoking.

In addition, you are to open the active line that gives the Food and Drug Administration's proposed regulations on tobacco by clicking on the underlined FDA entry. If you have trouble, you can use the following URL for the FDA site. You are supposed to record your opinion of the FDA proposals on tobacco regulation and give a reason for your opinion.

http://www.fda.gov/opacom/campaigns/tobacco.html

© 1999 Harcourt Brace & Company. All rights reserved.

Bridging the Gap   **II**   Name _____

Internet access to the Center for Disease Control, CDC and the Food and Drug Administration, FDA

1. Use the following URL to answer these questions.
   http://www.cdc.gov/nccdphp/osh/issue.htm

   What is the Center for Disease Control's estimate of the number of deaths in the United States that are annually caused by tobacco? Does this seem like a large number to you? Why?

2. Use the following URL to answer this question.
   http://www.hhs.gov/news/press/1996pres/960823b.htm

   What is your opinion of the regulations proposed by the FDA? Would you endorse these proposals? Explain.

3. Give the URL for a interesting site you found while doing this exercise.

   http://_____

   What makes this site noteworthy?

Bridging the Gap   III   Name_____
Internet access to the Environmental Protection Agency, EPA
Health Risk: EPA Office of Air and Radiation, OAR

The issue of health risk and air pollution concerns all of us. Frequently people assume that air pollution is only a problem in big cities like New York city or Los Angeles. It may come as a surprise, but even cities in the Rocky Mountains have air pollution problems. The Environmental Protection Agency, EPA, provides some basic information about risks at the URL given in this exercise.

The EPA page on Air Pollution and Health Risk
The EPA page on Air Pollution and Health Risk can be reached using the following uniform resource locator (URL).

    http://www.epa.gov/oar/oaqps/air_risc/3_90_022.html

You can find answers to the following questions at this URL.

1. What are the attributes of risks that are classed as more serious and of greatest concern?

2. What are the attributes of risks that are classed as less serious and of lesser concern?

3. According to the EPA Health Risk is the probability, or chance that exposure to a hazardous substance will make you sick. What are the two factors that are used to estimate risk?

4. What actions does the EPA recommend to reduce your risk of exposure to hazards?

© 1999 Harcourt Brace & Company. All rights reserved.

2 The Chemical View of Matter

2.1 Elements -- the Most Simple Kind of Matter

This section describes matter and the definitions for elements, pure substances and atoms. A *pure substance* is a material that is uniform and has a fixed composition at the submicroscopic level. An *element* is a pure substance made up of only one kind of atom like helium gas or pure gold metal. An *atom* is the smallest particle of an element that retains the characteristics of the element. The atoms of every element are unique for that element.

Objectives
After studying this section a student should be able to:
 describe the characteristics that distinguish elements from other substances
 describe the type of particles that make up an element
 write the definitions for a pure substance, element, and atom

Key terms
   atoms   elements   pure substances

2.2 Chemical Compounds -- Atoms in Combination

This section explains how to read chemical formulas. It tells what formulas represent and gives a definition of chemical compounds. The formula, H_2O, for water is interpreted. The subscript of "2" on the hydrogen tells the actual count of hydrogen atoms in the molecule of water. The subscript "1" on the oxygen atom tells the count of oxygen atoms in the molecule.

Ball and stick model of a water molecule

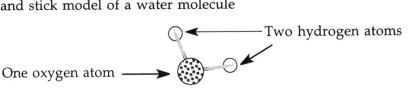

Space filling model of a water molecule

Objectives
After studying this section a student should be able to:
   use the chemical formula to tell what elements make up a compound
   tell the numbers of atoms of each element in a chemical formula
   describe the type of particles that make up a compound

Key terms
   chemical compound   molecule   subscript
   characteristic property chemical formula

2.3 Mixtures and Pure Substances

This section points out that most naturally occurring materials are mixtures. Homogeneous mixtures are uniform throughout like a well-mixed solution of salt and water. Heterogeneous mixtures, like concrete, have pockets of one substance mixed with pockets of a different substance. Mixtures are physical combinations. They can purified by mechanical means.

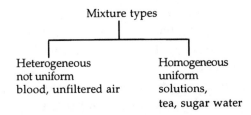

© 1999 Harcourt Brace & Company. All rights reserved.

Objectives
After studying this section a student should be able to:
- name the types of mixtures and give examples of the two types
- tell how homogeneous mixtures and heterogeneous mixtures differ
- describe what happens to a mixture when it is purified
- explain how purification of a mixture leads to a change in properties

Key terms
mixtures   heterogeneous mixtures   homogeneous mixtures
solutions   separation of mixtures   purification

2.4 Changes in Matter: Is It Physical or Chemical

This section describes physical properties, physical change, chemical properties, chemical change, energy, potential energy, kinetic energy, chemical reactions, and molecules. Physical properties are those that are associated with changes that do not alter the chemical identity of a substance. Examples of physical properties are boiling point, color, density, odor, melting point, conductivity of heat and electricity. Density is the ratio of mass to volume. Units for density are grams/mL or kilograms /Liter. When water boils, it undergoes a physical change from liquid state to gas state. The normal boiling point of water (a physical property) is 100°C.

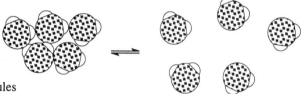

Water molecules in the liquid are in contact and have the formula, H_2O.

Water molecules in the gas state are far apart but still have formula H_2O.

Chemical change is a process that alters the chemical identity of a substance. Chemical properties describe how a substance reacts or behaves when mixed with another substance. A chemical reaction is a process in which one or more substances(reactants) are converted to one or more different substances (products). An example is the formation of water when hydrogen, H_2, and oxygen, O_2, react.

Reactants   Products
Hydrogen   Oxygen   Water

The reactants are changed into products. The smallest unit of many compounds like water is a molecule. Even the elements hydrogen and oxygen are found in nature as diatomic, two atom, molecules. Notice that 4 H atoms and 2 O atoms are needed to form the 2 water molecules. Energy changes are connected with physical and chemical change. Energy is defined as the ability to do work or produce change. Energy can be transferred as mechanical energy, heat, light, and electricity. Energy in storage is potential energy and energy associated with motion is kinetic energy. The total energy of an object is the sum of its potential and kinetic energies.

Objectives
After studying this section a student should be able to:
- identify and describe physical properties like boiling point, density, melting point
- describe chemical properties
- identify reactants and products if given a chemical reaction
- distinguish between physical and chemical properties
- give the definitions for energy, potential energy and kinetic energy

© 1999 Harcourt Brace & Company. All rights reserved.

Key terms

physical property	physical change	chemical property
density	molecule	chemical reaction
reactants	products	energy
kinetic energy	potential energy	forms of energy

2.5 Classification of Matter

This section gives more detailed descriptions of homogeneous mixtures, heterogeneous mixtures, pure substances. Major reasons for studying pure substances, and the connection between the structure of matter and chemical and physical properties are included.

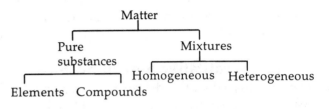

Objectives
After studying this section a student should be able to:
- describe the structure and properties of a homogeneous mixture
- describe the structure and properties of a heterogeneous mixture
- give reasons for studying pure substances

Key terms

heterogeneous	homogeneous	structure of matter

2.6 The Chemical Elements

This section includes a discussion of the classes of elements and their symbolic representation. Metals are described as lustrous, typically solids, good conductors of heat and electricity. Nonmetals are described as not lustrous, poor conductors of heat and electricity. Many nonmetals exist as diatomic molecules. The space filling models below are to scale.

Hydrogen, H_2 Nitrogen, N_2 Oxygen, O_2 Chlorine, Cl_2

Fluorine, F_2 Bromine, Br_2 Iodine, I_2

Objectives
After studying this section a student should be able to:
- describe what is meant by naturally occurring elements
- describe what is meant by the term diatomic molecule when referring to elements
- describe the properties of metals and nonmetals and identify common ones

Key terms

naturally occurring elements	nonmetals	metals
diatomic molecules	symbol	common nonmetals

2.7 Using Chemical Symbols

This section describes types of chemical formulae, use of subscripts, characteristics of organic and inorganic compounds, information available from chemical equations, and how to balance chemical equations.

Objectives
After studying this section a student should be able to:
- use the chemical formula to count the relative numbers of atoms of each element
- tell how structural formulas, molecular formulas, and condensed formulas differ
- explain what information is given by a structural formula
- state the definitions for organic and inorganic compounds

© 1999 Harcourt Brace & Company. All rights reserved.

describe the information provided by a chemical equation
identify reactants and products in a chemical equation
balance an unbalanced chemical equation
distinguish between coefficients in a chemical equation and formula subscripts

Key terms

chemical formulas	subscripts	structural formulas
molecular formulas	condensed formulas	organic compounds
inorganic compounds	chemical equations	balanced equations
coefficient	aqueous solution	

2.8 The Quantitative Side of Science

This section describes the difference between qualitative and quantitative information. For example measuring the mass of a lead brick on a scale is quantitative while lifting it with your hand and deciding that it is heavy is qualitative. The section discusses the metric system, fundamental units, derived units, prefixes in the metric system, and techniques used in converting one metric unit to another. It shows how to use conversion factors.

Objectives

After studying this section a student should be able to:
  tell how qualitative and quantitative information differ and give an example of each
  distinguish between qualitative and quantitative procedures and data
  give the names and symbols for the fundamental SI units of length, mass, time
  tell how a fundamental unit differs from a derived unit and give an example of each
  name common metric prefixes, give their abbreviation and meaning; for example

   milli-, m, $\frac{1}{1000} = 0.001 = 10^{-3}$

  give symbols for common units such as meter, liter, gram
  use the unit conversion method to convert one metric unit into another
  use the unit conversion method to convert a metric unit to an English or American unit

Key terms

qualitative	quantitative	metric system
fundamental units	derived units	physical quantity
length	amount of substance	time
meter = m	prefixes	mega = M = 1,000,000
kilo = k = 1000	centi = c = $\frac{1}{100}$ = 0.01	milli = m = $\frac{1}{1000}$ = 0.001
micro = µ = 0.000001	liter = L	gram = g

Additional readings

Hoffman, Roald. "How Should Chemists Think?" <u>Scientific American</u> Feb. 1993: 66.

Itano, Wayne M. and Norman F. Ramsey. "Accurate Measurement of Time." <u>Scientific American</u> July 1993: 56.

Lipkin, Richard. "Physicists Spot Element 111." <u>Science News</u> 7 Jan. 1995: 5.

Nauenberg, Michael et al. "The Classical Limit of an Atom." <u>Scientific American</u> June. 1994: 44.

Nelson, Robert A. "Guide for Metric Practice." <u>Physics Today</u> Aug. 1995: 15.

Raloff, Janet.. "Microwaves Accelerate Chemical Extractions." <u>Science News</u> 21 Aug. 1993: 118.

Seligman, Daniel. "Metric mania." <u>Fortune</u> 19 Oct. 1992: 132..

Zebrowski, George. "Time is nothing but a clock." <u>Omni</u> Oct. 1994: 80.

Answers to Odd Numbered Questions for Review and Thought

1. Materials typically used in an average day that were not chemically changed from their natural states are gold, oxygen in the air, water, methane in natural gas, beeswax, wood.

3. No. Two pure substances that have the same set of physical and chemical properties would be the same. Two different substances will have different formulas or structures. This typically produces different physical and chemical properties. Some properties may be the same, but it is highly unlikely that all would be the same.

5. When a BIC lighter is lit, the spark provides the energy needed to raise the temperature of the air and BIC fuel mixture above the ignition temperature. The mixture ignites and bursts into flame. The mixture of oxygen, $O_{2(gas)}$ and hydrocarbon fuel are chemically changed to heat and new substances water vapor, $H_2O_{(gas)}$, and carbon dioxide, $CO_{2(g)}$.

7. a. Density is a physical property because it depends only on the physical properties of mass and volume. The mathematical definition for density is d = mass/ volume
    b. Melting temperature is a physical property. It is linked to the amount of energy needed to separate the particles in a solid and allow them to move freely in the liquid. The particles stay intact, but are no longer tied down to definite locations in the solid.
    c. Decomposition of a substance into two elements upon heating is a chemical property because the substance changes from its original formula to two elements. There is a change in chemical composition. The decomposition breaks the larger structure into the elements that combined to form the substance. An example is water, it can be decomposed into hydrogen and oxygen.
    d. Electrical conductivity is a physical property. The chemical composition of a piece of copper wire does not change when it conducts electricity.
    e. The failure of a substance to react with sulfur is a chemical property. The chemical composition is unchanged when no reaction occurs.
    f. The ignition temperature of a piece of paper is a chemical property. It indicates the ease with which the paper reacts with air resulting in a change in composition.

9. a. Mercury is an element. Division of a sample of mercury leads to separate Hg atoms.
    b. Milk is a mixture of water, minerals, proteins, fats.
    c. Pure water is a compound, contains only one kind of molecule (H_2O)
    d. Wood is a mixture of cellulose, water. Wood changes weight when dried and loses H_2O.
    e. Ink is mixture of dye and solvent. There is more than one type of molecule in mixture.
    f. Iced tea is a mixture of water, caffeine, tea extract. More than one type of molecule.
    g. Pure ice is a compound, solid pure water containing only one kind of molecule, (H_2O).
    h. Carbon contains one kind of atom. It is an element, it contains only C atoms.
    i. Antimony contains one kind of atom. It is an element, it contains only Sb atoms.

© 1999 Harcourt Brace & Company. All rights reserved.

11. Properties of iron do not change because all particles in iron are atoms of iron. Steel is an alloy or mixture of iron and other atoms. The type of steel depends on what is added. Each type of steel has a different set of properties. Steels differ in hardness, reactivity, and ability to be drawn into wire.

13.

	Major Source	Compound
nitrogen	air	ammonia, NH_3
sulfur	underground deposits	sulfuric acid, H_2SO_4
chlorine	sea water	sodium chloride, NaCl
magnesium	sea water	Milk of Magnesia, $Mg(OH)_2$
cobalt	mineral deposits	cyanocobalamin, Vitamin B_{12}

15. Cytoxan has the formula $C_7H_{15}O_2N_2PCl_2$.
    a. The number of atoms in the formula equal the total of the subscripts; twenty nine, 29.
       7 + 15 + 2 + 2 + 1 + 2 = 29
    b. There are six different elements in cytoxan; carbon, C; hydrogen, H; oxygen, O; nitrogen, N; phosphorus, P; and chlorine, Cl.
    c. The ratio of hydrogen atoms to nitrogen atoms in cytoxan is 15/2.
    d. Yes, cytoxan is considered to be an organic compound because it has a carbon based structure.

17. The three states of matter, solid liquid and gas, are familiar to all of us. It is possible to convert most any substance from one form to another provided it doesn't decompose during the heating process. Wood will burst into flame if heated in air and it will decompose if heated in an airless container. Usually you can melt a solid and then heat the liquid hot enough to boil it. Ice, solid water, can be melted and then boiled in this way.

19. Yes, a mixture of H_2 and O_2 can exist at room temperature. This mixture will be stable as long as no spark or energy is added to the gas mixture. A reaction produces water, H_2O, which contains both elements. The properties of the elements are no longer observed after the water is formed.

21. a. On the left side of the arrow, "2 Na" means 2 Na atoms; one Cl_2 molecule contains 2 Cl atoms. On the right side 2 NaCl units contain 2 Na atoms and 2 Cl atoms.

    Two sodium atoms react with   Yield   two formula units of
    one chlorine molecule   sodium chloride

    b. On the left one N_2 molecule contains 2 N atoms and 3 Cl_2 molecules contain 6 Cl atoms. On the right 2 NCl_3 molecules contain a total of 2 N atoms and 6 Cl atoms.

c. On the left there are 1 C atom, 2 H atoms and 2 + 1 = 3 O atoms. On the right there are 1 C atom, 2 H atoms and 3 O atoms.

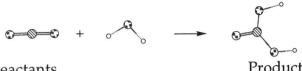

Reactants   Products

d. On the left there are 4 H atoms and 4 O atoms in 2 molecules of H$_2$O$_2$. On the right there are 4 H atoms in the 2 molecules of water; there are also 2 O atoms in the 2 water molecules and 2 more O atoms in the O$_2$ molecule for a total of 4 O atoms.

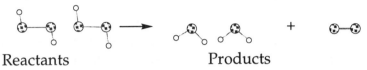

Reactants   Products

23. a. On the left side of the arrow, "2 Na" means 2 Na atoms; one Cl$_2$ molecule contains 2 Cl atoms. On the right side 2 NaCl units contain 2 Na atoms and 2 Cl atoms.

Reactants   Products

b. On the reactants side or left, one N$_2$ molecule contains 2 N atoms and three Cl$_2$ molecules contain 6 Cl atoms. On the right two NCl$_3$ molecules contain a total of 2 N atoms and 6 Cl atoms.

Reactants   Products

c. On the left there are 1 C atom, 2 H atoms and 2 + 1 = 3 O atoms. On the right there are 1 C atom, 2 H atoms and 3 O atoms.

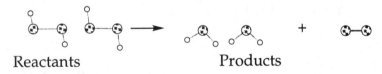

Reactants   Products

d. On the left there are 4 H atoms and 4 O atoms in 2 molecules of H$_2$O$_2$. On the right there are 4 H atoms in the 2 molecules of water; there are also 2 O atoms in the 2 water molecules and 2 more O atoms in the O$_2$ molecule for a total of 4 O atoms.

Reactants   Products

25. The tea in tea bags is a mixture. It can be partially separated by dissolving some water-soluble substances with hot water. Instant tea is a mixture of the water-soluble substances in tea.

27. a. 1 gram = 1000 milligrams
    b. 1 kilometer = 1000 meters
    c. 1 gram = 100 centigrams

© 1999 Harcourt Brace & Company. All rights reserved.

29. A density unit is literally a measure of mass per unit volume. This means the numerator must be grams or a related unit and the denominator must be milliliters or another volume unit.
    a. 9 cal/gram; no.
    b. 100 cm/meter; no.
    c. 1.5 g/mL; yes. Grams/milliliter is mass/volume.
    d. 454 g/ lb.; no.

Solutions for Problems

1. The answer is supposed to be in units of acres per cow. The relation between acres and cows is 1 acre/10 cows. This means there would be 5 acres/50 cows or 6 acres/60 cows. Multiply top and bottom by 5.5.

    $$\frac{? \text{ acres}}{55 \text{ cows}} = \frac{1 \text{ acre}}{10 \text{ cows}} \times \frac{5.5}{5.5} = \frac{5.5 \text{ acres}}{55 \text{ cows}} \text{ or } 5.5 \text{ acres}/55 \text{ cows}$$

2. An Advil® tablet = 200 mg ibuprofen. Converting the mg to grams uses the link between grams and milligrams. 1000 mg = 1 g

    $$\frac{200 \text{ mg ibuprofen}}{1 \text{ tablet}} \times \frac{1 \text{ gram}}{1000 \text{ mg}} = \frac{0.200 \text{ gram ibuprofen}}{1 \text{ tablet}}$$

    The conversion of micrograms to grams is needed for the second part. 1,000,000 µg = 1 g

    $$\frac{200 \text{ mg ibuprofen}}{1 \text{ tablet}} \times \frac{1 \text{ gram}}{1000 \text{ mg}} \times \frac{1,000,000 \text{ µg}}{1 \text{ gram}} = \frac{200,000 \text{ µg ibuprofen}}{1 \text{ tablet}}$$

3. The answer is 10,000 meters. This is determined by using the definition of kilometer in terms of meters, 1 km = 1000 m. The appropriate conversion factor is $\frac{1000 \text{ meters}}{1 \text{ km}}$

    $$? \text{ meters} = 10 \text{ km} \times \frac{1000 \text{ m}}{1 \text{ km}} = 10,000 \text{ meters}$$

4. The answer is 3000 mg protein / 1 oz. cereal. The conversion of grams to milligrams must be done. The appropriate conversion factor is $\frac{1000 \text{ mg}}{1 \text{ g}}$

    $$\frac{? \text{ mg}}{1 \text{ oz}} = \frac{3.00 \text{ g protein}}{1 \text{ oz}} \times \frac{1000 \text{ mg}}{1 \text{ g}} = \frac{3000 \text{ mg protein}}{1 \text{ oz cereal}}$$

5. a. Use the definition 100 cm = 1 m as a conversion factor; $0.04 \text{ m} = 4 \text{ cm} \times \frac{1 \text{ meter}}{100 \text{ cm}}$

    b. Use the conversion factor 1000 mg = 1 g; $43 \text{ mg} = 0.043 \text{ g} \times \frac{1000 \text{ mg}}{1 \text{ g}}$

    c. Use the conversion factor 1000 mm = 1 m; $15500 \text{ mm} = 15.5 \text{ m} \times \frac{1000 \text{ mm}}{1 \text{ meter}}$

    d. Use the conversion factor 1000 mL = 1 L ; $0.328 \text{ L} = 328 \text{ mL} \times \frac{1 \text{ L}}{1000 \text{ ml}}$

    e. Use the conversion factor 1000 g = 1 kg ; $980 \text{ g} = 0.980 \text{ kg} \times \frac{1000 \text{ g}}{1 \text{ kg}}$

6. The gorilla's mass needs to be converted from kilograms to grams. The kilogram to gram relationship is, 1 kilogram = 1000 grams. The gorilla's mass is

$$163 \text{ kg} \times \frac{1000 \text{ g}}{1 \text{ kg}} = 163000 \text{ g}.$$

7. The average man's mass in kilograms is converted to milligrams using the links between grams and kilograms and the milligram/gram conversion factor.
1 kg = 1000 grams   1 gram = 1000 milligrams

$$70 \text{ kg} \times \frac{1000 \text{ g}}{1 \text{ kg}} \times \frac{1000 \text{ mg}}{1 \text{ g}} = 70,000,000 \text{ mg}$$

© 1999 Harcourt Brace & Company. All rights reserved.

Speaking of Chemistry

The Chemical View of Matter

Name _____

Across
5 Means dissolved in water (aq)
6 Abbreviation for centimeter
8 Property with units of mass/unit volume
9 Measured in kilograms and grams
10 Prefix meaning multiply by 1000
12 Made in a chemical reaction
13 Element with symbol "C"
15 A nonuniform mixture
19 Prefix for one million
21 Matter with no definite volume or shape
22 Inventor of dynamite

Down
1 Prefix that means 1/1000
2 Equations with equal numbers of atoms of each element on both sides
3 Elements and compounds are two forms of _____ substances.
4 Element with Latin name, aurum.
6 Prefix meaning 1/100
7 Volume unit equal to 1000 mL
9 SI unit that equals 39.4 inches
11 Measured in units of meters
12 Property like density and color
14 Smallest particle of an element
15 Element with symbol "H"
16 Measured in units of seconds or minutes
17 Measured in calories
18 Element with symbol "S"
20 Mass equal to 1/1000 kilogram.

© 1999 Harcourt Brace & Company. All rights reserved.

Bridging the Gap   I
Separation of Mixtures, Paper Chromatography

Mixtures are either homogeneous or heterogeneous. Homogeneous mixtures are uniform in composition. Heterogeneous mixtures are not. Salt water is a solution of water and NaCl and is homogeneous if thoroughly mixed. Both types of mixtures can be separated into by physical means. Food colorings are typically a homogeneous mixture of a solvent a single dye or a combination of selected dyes that produce the desired color. You will use paper chromatography to test a food coloring to see if it is a pure substance of a single dye or mixture of dyes and solvent. Other substances that are not colored can be detected using ultraviolet or black light. These substances appear to glow in the dark. Your exercise is simpler but uses very essential principles.

Equipment and materials
Clear colorless glass or plastic tumbler or jelly jar, paper coffee filter, toothpicks, scissors, pencil, adhesive tape, tap water and a package of Schilling® or other brand of food coloring.

1. Cut a half inch wide (1.25 cm) strip of coffee filter paper about four inches long(10 cm).

2. Make a start line with pencil mark at half an inch from one end that will be the bottom.

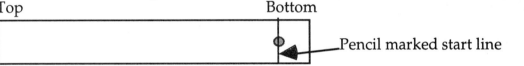

3. Cut off one end of a toothpick. Dip the fresh cut end into the food coloring.

4. Use the toothpick to place a dot of blue food coloring on the pencil mark start line and allow it to dry.

5. Attach a piece of tape to the top end of the strip of paper. Tape the paper to the pencil and lower the paper into the tumbler. Check how far the paper projects into the container. Remove the pencil and paper.

6 Add water to container so the water level will touch the paper 1/2 inch below the start line.

7. Lower the paper into the tumbler so the water touches the end of the paper at least 1/2 inch below the spot.

8. Let the water wick up the paper. Note where the water wets the paper, this is the solvent front. The water will climb the first inch quickly. The dye will trail behind the water.

9. When the front edge of the water reaches half way up the paper remove the paper from the glass. Allow the paper to dry. Note if more than one color appears on the paper.

10. Repeat steps 1 through 8 but using the yellow dye and then the green dye. Record your observations on the report sheet. Which of the dyes are single colors and which are mixtures. Explain your reasoning.

© 1999 Harcourt Brace & Company. All rights reserved.

Bridging the Gap I Name _____
Separation of Mixtures, Paper Chromatography

Data

Food coloring	Colors observed
Blue	
Yellow	
Red	
Green	

Food coloring	Mixture Yes/ No	How many components are in the dye.
Yellow		
Blue		
Red		
Green		

Concepts and analysis

Which of the food coloring dyes is a mixture? Explain your reasoning.

Do you think that other substances like vegetable dyes or inks could be tested using this method? Explain.

Explain briefly what would have to be done if the dyes would not dissolve in water? .

Name an everyday activity that involves the separation of a mixture. For example tea is brewed using hot water to extract desirable substances from the tea leaves.

Bridging the Gap   II   Name _____
Internet access to the National Institute of Standards and Technology
"To be or not to be": the English or the Metric system

The English system of measurement used today in the United States originated in the decrees of English monarchs. The French Revolution produced the overthrow of the French monarchy and in 1799 it also led to the creation of the set of weights and measures we call the metric system. The metric system was legalized for use in the United States in 1866 along with the traditional English system. Today the only countries in the world that do not use the metric system are the United States of America, Liberia and Myanmar.

United States government policy toward the Metric System
This part is to access the following URLs for the National Institute of Standards and Technology, NIST, and review the evolution of the relationship between the United States and the Metric System. You can answer the first question after reviewing this NIST page.

> http://ts.nist.gov/ts/htdocs/200/202/ic1136a.htm

1. What was the year when the United States signed the "Treaty of the Meter?

2. Access this FDA page to answer the following question.

    > http://www.fda.gov/ora/inspect_ref/itg/itg30.html

    What does the term "both timeless and toothless" have to do with the Metric Conversion Act of 1975? In your opinion does this influence the pace of metrification in the United States ?

Another part of your assignment is to write a brief argument on the back of this page for adopting the metric system and replacing the English system. Lastly you are to write an argument for continuing the current pattern of using two systems.

© 1999 Harcourt Brace & Company. All rights reserved.

3. Argument for adopting the Metric system and dropping the English system

4. Argument for retaining both the English system and the Metric system

Bridging the Gap   III   Name _____
Internet access to the National Institute of Standards and Technology "A Walk Through Time"

Time measurement has been of tremendous importance through the ages. The time to plant crops to avoid frosts was a matter of survival for cultures that tilled the soil. Ancient people needed to know when seasonal changes would occur to protect themselves just as we do today. People needed to know when the monsoon or seasonal flood was imminent. The NIST page titled "A Walk Through Time" provides an excellent review of time measurement from the days of prehistory through to the "atomic clocks" of today. This site is well done and is rich with information.

Ancient time measurement

There are a series of linked pages each with a different URL. You can answer the following questions after reviewing the NIST sites.

        http://physics.nist.gov/GenInt/Time/time.html

        http://physics.nist.gov/GenInt/Time/ancient.html

1. When did the Egyptians develop the first 365-day calendar?

2. The Mayans of Central America relied on three celestial bodies to establish their 260-day and 365-day calendars. What were these celestial bodies?

A Revolution in Timekeeping

        http://physics.nist.gov/GenInt/Time/revol.html

3. Spring powered clocks were invented between 1500 and 1510 by Peter Henlein of Nuremberg. What is the advantage of a spring powered clock over a water clock or pendulum?

© 1999 Harcourt Brace & Company. All rights reserved.

4. In the section on the Quartz Clock was developed in the 1930s and 1940s. Why are quartz clocks better than pendulum and balance wheel clocks?

The "Atomic Age" of Time Standards

   http://physics.nist.gov/GenInt/Time/atomic.html

5. The atomic clock was formally recognized as the new international unit of time in 1967. What is the atom used as the reference for the second in the atomic clock?

3 Atoms and the Periodic Table

The Greek Influence on Atomic Theory
The main topic in this section is the historical development of the theory of matter. Both the continuous theory proposed by Plato (427-347 BC.) and Aristotle (384-322 BC.) and the atomic theory proposed by Democritus and Leucippus, and later refined by John Dalton in 1803, are presented.

Objectives
After studying this section a student should be able to:
- describe the continuous view of matter proposed by Plato and Aristotle
- describe the view of matter proposed by Democritus and Leucippus
- give the time periods when Democritus and John Dalton proposed their theories

Key terms
atomic theory	Democritus	Leucippus
Dalton	Plato	Aristotle

3.1 John Dalton's Atomic Theory
A picture of John Dalton's life (1766-1844) in England is detailed in this section. Dalton proposed his atomic theory in 1803. His theory was consistent with the law of conservation of matter, and the law of definite proportions which are discussed in this section. Dalton's theory rests on these postulates:
1. All matter is made up of indestructible atoms.
2. Atoms of the same element are identical.
3. Compounds result from the combination of atoms of different elements in ratios of small whole numbers.
4. Definite arrangements of atoms exist in elements and compounds. Chemical change occurs when these atomic combinations are rearranged.

The law of conservation of matter holds that matter is neither created nor destroyed. It can be transformed by physical change from solid to liquid to gas and back again. If you weigh an ice cube when it is a solid and then when it melts to form liquid water, the weights will be the same. The law of definite proportions results from the observation that all pure substances always have constant proportions by mass for each constituent. The salt sodium chloride, NaCl, is always 39.3 parts sodium and 60.7 parts chlorine by mass. This means 100 pounds of NaCl will contain 39.3 pounds of sodium and 60.7 pounds of chlorine. Pure water, H_2O, is always 11.2% hydrogen and 88.8% oxygen by mass.

Objectives
After studying this section a student should be able to:
- describe and state the four provisions of Dalton's atomic theory
- state two scientific laws that Dalton's atomic theory attempted to explain
- state and describe the law of conservation of matter
- state and describe the law of definite proportions

Key terms
Dalton	Proust	atoms
law of conservation of matter	Lavoisier	law of definite proportions

© 1999 Harcourt Brace & Company. All rights reserved.

3.2 Structure of the Atom

This section deals with the origins of radioactivity, descriptions of historic experiments that revealed the properties of subatomic particles, the roles of scientists like Becquerel, Curie, Marsden, Geiger, and Rutherford. It discusses the properties of electrons, protons, and neutrons. It describes the properties of radioactive radiation like beta (β^{1-}) rays, alpha (α^{2+}) rays, and gamma (γ) rays. Rutherford (1852-1908) proposed the nuclear model for the atom based on his alpha particle scattering experiments. Rutherford proposed that the atom size is 100,000 times greater than the atom nuclear core size.

Objectives
After studying this section a student should be able to:
- tell how a radioactive element differs from a nonradioactive one
- describe Becquerel's and Marie Curie's role in the study of radioactivity
- describe how charged particles interact
- name and describe the three subatomic particles, their charge, mass and location in an atom
- describe the physical properties of alpha particles, beta particles and gamma rays
- sketch and label the apparatus used in Rutherford's gold foil experiment
- describe the observations and interpretation of Rutherford's gold foil experiment
- describe the relative sizes of the nucleus and the atom

Key terms

natural radioactivity	Marie Curie	Becquerel
atomic structure	electrical charge	Rutherford's experiment
electrons	neutrons	protons
electrically neutral	nucleus	Rutherford's model
beta particle, β	gamma rays, γ	alpha particle, α

3.3 Modern View of the Atom

This section describes the locations of subatomic particles in the atom, the methods to measure positions of atoms with the scanning tunneling microscope (STM), and the atomic force microscope (AFM). Scientific notation is used to describe the relative size of atoms.

The relation of atomic number and mass number to atomic structure is explained. The format used to describe the number of particles in atoms is described.

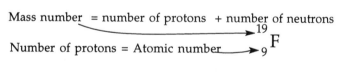

Isotopes, their relative abundances, atomic weights, the atomic mass unit, average atomic weight are described. An introduction to reading periodic table entries is given.

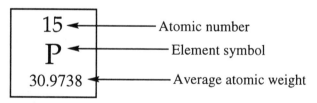

Objectives
After studying this section a student should be able to:
- tell what STM and ATM are used for in the study of atoms
- tell how the number of electrons and the number of protons are related in a neutral atom
- describe the locations for electrons, protons and neutrons in an atom
- give the definitions for atomic number, mass number and isotopes
- explain how to determine the neutron number from mass number and atomic number

identify a pair of isotopes from information and notation like $^{21}_{10}Ne$, $^{22}_{10}Ne$, $^{23}_{11}Na$, $^{19}_{9}F$.
tell how the atomic mass unit is defined
describe what abundance means when describing isotopes and average atomic weight
read the periodic table to identify an element with a specific atomic number

Key terms

atomic number, Z	scanning tunneling microscope	STM
mass number, A	atomic force microscope	AFM
isotopes	atomic mass unit, amu	carbon-12
relative abundance	atomic weight	

3.4 Where Are the Electrons in Atoms?

This section includes descriptions of emission line spectra, continuous spectra, the interaction between a light beam and a prism, the relationship between wavelength, color, frequency and energy. The Bohr model for the hydrogen atom and the idea of quantized energy levels are explained. Quantum theory, ground states, and excited states are introduced. Examples of electron arrangements for atoms, valence electrons, and Lewis dot formulas are explained and illustrated with examples.

Objectives

After studying this section a student should be able to:
tell how an emission line spectrum is produced
tell how an emission line spectrum differs from a continuous spectrum
describe how wavelength, energy and frequency are related to the colors of visible light
describe what is meant by an energy level and state the restrictions on the values for "n"
explain how the ground state for an atom relates to an excited state
write out the ground state Bohr electron arrangement for atoms with atomic number 1-20
predict the limit on the electron population for an energy level using the $2n^2$ rule
identify valence electrons in an atom and describe their importance
match a Bohr electron arrangement like 2-2-1 to the correct element, boron

Key terms

emission line spectrum	continuous spectrum	electromagnetic spectrum
visible light	white light	infrared radiation
X-rays	ultraviolet radiation	quantum
frequency, ν	wavelength, λ	$\nu \lambda = c$
Bohr theory	energy levels or shells	principal quantum number
ground state	excited state	emitted light
$2n^2$	valence electrons	Bohr electron arrangement

3.5 Development of the Periodic Table

This section describes Mendeleev's (1834-1907) and Meyer's work and their proposal of the periodic law and a periodic arrangement of the 69 then known elements based on atomic mass. Mendeleev's prediction of the existence of "missing" elements and their properties was a success of his theory. Mosley's X-ray experiments that linked periodic properties to atomic number are also discussed. The placing of elements with similar chemical properties in the same group is explained. Periods are defined as the horizontal rows in the periodic table.

Objectives

After studying this section a student should be able to:
give the definitions for atomic mass and atomic number
describe Mendeleev's contribution to the development of the periodic table

describe the historical development of the periodic table and state the periodic law
state the atomic property that governs the position of an element in the periodic table
explain why Mendeleev was able to predict the properties of undiscovered elements

Key terms

periodic trend   trend in properties   periodic law
increasing atomic number Mendeleev   Moseley

3.6   The Modern Periodic Table

This section gives the reasons for the classification of elements. The definitions are given and explained for period, A groups, B groups, main-group elements, transition elements, inner transition elements, lanthanide and actinide series. It describes the physical properties that are used to classify elements as metals, nonmetals, and metalloids. Examples of elements in these classifications are given. The groups for semi-conductors and noble gases are identified.

Objectives

After studying this section a student should be able to:
describe the significance of a group in the periodic table
locate the main-group, transition and innertransition elements
tell how main-group elements differ from the rest of the periodic table
tell the group of an element if given the element symbol and a periodic table
describe the physical properties of metals, nonmetals and metalloids
explain why the periods of the main group elements are typically eight elements long
locate metals, nonmetals, noble gases, semiconductors, and metalloids in a periodic table
locate A group and B group elements when given a periodic table

Key terms

representative groups   main-group   semiconductors
B groups   transition elements   innertransition elements
metals   malleability   periods
nonmetals   metalloids   ductility
noble gases   insulators

3.7   Periodic Trends

This section describes the Lewis dot symbol notation for atoms and ions. It shows how the Lewis dot symbols for atoms change when the atoms lose electrons to form cations and how the symbols change when the atoms gain electrons to form anions. The section includes an explanation of the importance of valence electrons. This section describes how metallic properties, atomic radii, and reactivity vary for elements in the same group. The variations of these properties for elements in a period are described and explained. The relationship between atomic size and reactivity is explained for metals and nonmetals.

Objectives

After studying this section a student should be able to:
give the definition for valence electrons
write the Lewis dot symbol for a representative element if given a periodic table
give the definition for a Lewis dot symbol
tell how group number and valence electrons relate
explain why valence electrons are chemically interesting
describe and explain how atomic radii vary for the members of a group
describe and explain the variation of atomic radii across a period
describe and explain how reactivity varies for the members of a group
relate reactivity to atomic radius

Key terms

- kernel
- valence electrons
- metallic character
- increase in reactivity
- outermost shell
- Lewis symbol
- atomic properties
- gain electrons
- reactivity
- dot symbol
- atomic radii (radius)
- lose electrons

3.8 Properties of Main-Group Elements

This section includes descriptions of the physical properties and the uses of some main group elements and their compounds. Formulas for oxides formed by alkaline earth metals (Be, Mg, Ca, Sr, Ba, and Ra) are given. Sources and specific applications of the alkali metals, alkaline earths are listed. Elements in Group IA, IIA, halogens and the noble gases are described.

Objectives

After studying this section a student should be able to:
- identify the representative elements using a periodic table
- tell which elements are metals, nonmetals and metalloids
- match elements with their group: halogens, noble gases, alkaline earth metals, etc.
- give an example of a compound formed by a typical member of each group
- tell why Group VIIIA elements are called the noble gases
- give the formula for the oxide compound formed by a Group IIA element i.e. M_xO_y
- give the formula for the halide compound formed by a main group element i.e. M_xX_y
- describe the general properties of members of each group discussed in the section
- name a common substance, if any exist, formed by an element from each group

Key terms

- group properties
- halogens
- salts
- alkali metals
- noble gases
- alkaline earth metals
- oxide

Additional readings

Binnig, Gerd, & Heinrich Rohrer. "The Scanning Tunneling Microscope" <u>Scientific American</u> Aug. 1985: 50.

Boslough, John. "Worlds within the Atom" <u>National Geographic</u> May 1985: 634.

Conkling, John A. "Pyrotechnics" <u>Scientific American</u> July 1990: 96.

Connes, Pierre. "How Light Is Analyzed" <u>Scientific American</u> Sept. 1968: 2.

Crommie, M. F. et al. "Confinement of Electrons to Quantum Corrals on a Metal Surface" <u>Science</u> 8 Oct. 1993: 218.

da C. Andrade, E.N. "The Birth of the Nuclear Atom" <u>Scientific American</u> Nov. 1956: 93.

Fowler, William A. "The Origin Of The Elements" <u>Scientific American</u> Sept. 1956: 82.

Nassau, Kurt "The Causes Of Color" <u>Scientific American</u> Oct. 1980: 124.

Perlman, I. & G. T. Seaborg. "The Synthetic Elements" <u>Scientific American</u> April 1950: 38.

Zare, Richard N. "Laser Separation Of Isotopes" <u>Scientific American</u> Feb. 1977: 86.

© 1999 Harcourt Brace & Co. All rights reserved.

Answers to Odd Numbered Questions for Review and Thought

1. Matter is neither created nor destroyed. Examples: flash bulb before and after use, hard boiling an egg, yarn knitted into clothing, melting ice cubes in a glass. There is no change in mass during the chemical changes (first two) or the physical changes.

3. Dalton knew that matter was conserved in chemical reactions. He knew that the composition of elements was always the same for a specific element. He also knew that pure substances always had the same definite composition. These facts were unknown to the early Greeks.

5. a. Rutherford used a beam of alpha particles, $^4_2\alpha$, from a radioactive source to bombard a thin gold foil. The alpha particles could be detected because they produce a pulse of light when they strike a phosphorescent screen. The foil target was positioned in the center of a circle with the alpha particles entering from one side. A phosphorescent coated "rim" surrounded the centrally positioned foil. When alpha particles passed through the foil they caused light flashes on the phosphorescent screen.

   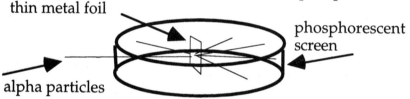

   b. Alpha particles usually passed through the foil, but occasionally they were reflected back towards the source.
   c. Rutherford proposed a planetary or nuclear model for the atom. Rutherford proposed that the alphas that passed straight through encountered electrons and could knock the light electrons out of the way. The reflected alphas were reflected back by a heavy, positively-charged nucleus.

7. a. The atomic number equals the number of protons in the nucleus of an atom.
   b. The mass number equals the total count of protons plus neutrons in the nucleus of an atom. Example $^{21}_{10}Ne$ has a mass number equal to 21, with 10 protons and 11 neutrons.
   c. The atomic weight equals the mass of an average atom of an element compared to an atom of ^{12}C which is assigned a mass of exactly 12 atomic mass units. For example magnesium, Mg-24, is an atom that is twice the mass of the carbon-12 reference.
   d. Isotopes are atoms with the same number of protons but with different numbers of neutrons. Example $^{21}_{10}Ne$ and $^{22}_{10}Ne$ have the same number of protons, 10, but different numbers of neutrons.
   e. The natural abundance on Earth, refers to the % that the isotope is of all the atoms of that element.
   f. The atomic mass unit, amu, is one-twelfth of the mass of the carbon-12 atom.

9. Elements of the same element all have the same atomic number. This means they have the same number of protons.

11. An atom of lithium always has 3 protons; the mass number 7 means that the sum of the numbers of protons and neutrons = 7, so it has 4 neutrons. The number of neutrons is determined by subtracting the atomic number from the mass number.
A = mass number; Z = atomic number
A - Z = number of neutrons; 7 - 3 = 4

13. Atoms with the same number of protons and different numbers of neutrons are isotopes like these isotopes of neon, $^{21}_{10}Ne$ and $^{22}_{10}Ne$. In the table A and D both have 25 protons but different numbers of neutrons so they are isotopes. B and C both have 24 protons but different numbers of neutrons so they are isotopes. Any pair of atoms with different numbers of protons are atoms of different elements.

15. The atomic number has the symbol Z. The number of protons equals the atomic number.
    a. Germanium, Z = 32. or 32 protons
    b. Silicon, Z = 14 or 14 protons
    c. Nickel, Z = 28 or 28 protons
    d. Cadmium, Z = 48 or 48 protons
    e. Iridium, Z = 77 or 78 protons

17. Yes, the statement is correct. The atomic mass of 24.305 is a weighted average based on the masses of individual Mg isotopes and their abundances. This means there are no atoms that actually weigh 24.305 amu.

19. Choice "c" is correct. The ratio by weight of the elements in a compound is constant. For example, samples of methane, CH_4, will always have the relative masses of 12 grams of carbon for every 4 grams of hydrogen. The atomic weights of carbon are 12.00 amu and for hydrogen 1.00 amu. The total weight for the atoms, CH_4, in a will add to 16.00 amu. The weight will always have these relative contributions to all samples of methane.

21. | Bohr arrangement | Number of electrons | Element |
    | --- | --- | --- |
    | 2-4 | 6 | Carbon |
    | 2-8 | 10 | Neon |
    | 2-8-3 | 13 | Aluminum |
    | 2-8-8-2 | 20 | Calcium |

    Notice that the first energy level can only accept 2 electrons and is always completely filled when the electron population is more than two. The second energy level has room for only eight electrons. When the electron population exceeds 10, the additional electrons must go into the third energy level.

23. Na, 1   Mg, 2   Al, 3   Si, 4   P, 5   S, 6   Cl, 7   Ar, 8
    The number of valence electrons can be predicted from the Group number. For example sodium, Na, is in Group I so sodium atoms have only one valence electron. The aluminum atom is in Group III and it has three valence electrons. The rule is useful mainly for the representative or main group elements.

25. When elements are arranged in order of their Atomic Numbers, their chemical and physical properties show repeatable trends.

© 1999 Harcourt Brace & Co. All rights reserved.

27. The known elements were arranged in order and gaps between elements occurred. These gaps were later filled in because the properties of the neighboring elements indicated the nature of the missing element, helping scientists discover the missing element.

31. a. The periods are the horizontal rows in the periodic table. There are 7.
    b. There are 8 representative groups. These are identified by a Roman numeral and a capital "A". The first representative group is IA. It includes H, Li, Na, K, Rb, Cs, and Fr. The last representative group is VIIIA. It includes He, Ne, Ar, Kr, Xe, Rn. These are the rare or noble gases.
    c. There are two representative groups that are all metals if you ignore H in IA. Group IA includes Li, Na, K, Rb, Cs, and Fr. The other is IIA. It includes Be, Mg, Ca, Sr, Ba, and Ra.
    d. The elements in Group VIIA and VIIIA are all nonmetals.
    e. Yes, all the elements in period 7 are metals. These are
    Fr Ra Ac Th Pa U   Np Pu Fm Cm Bk Cf
    Es Fm Md No Lr Rf Ha Sg Ns Hs Mt

33. All elements in the same group have the same number of valence electrons and have similar chemical properties. For example all of the elements in Group IA have one valence electron and react to form positive ions with a plus one charge.
    Choices "a. 2,1" and "g. 2, 8, 8, 1" both have filled inner levels and one electron in the outermost level.
    Choices "b. 2,6" and "d. 2,8,6" both have filled inner levels and have 6 electrons in the outermost level.

35. a. All of the elements in Group IIIA have three valence electrons.
        B, Al, Ga, In, Tl each have 3 valence electrons
    b. All of the elements in Group IVA have four valence electrons.
        C, Si, Ge, Sn, Pb all have 4 valence electrons
    c. All of the elements in Group VIIA have seven valence electrons.
        F, Cl, Br, I, At all have 7 valence electrons
    d. All of the elements in Group IA have one valence electron.
        H, Li, Na, K, Rb, Cs, Fr all have 1 valence electron

37. The elements in the same group are more metallic at the bottom of the group. Element to the left of the periodic table are more metallic.
    a. Be is more metallic than B
    b. Ca is more metallic than Be
    c. Ge is more metallic than As
    d Bi is more metallic than As

39. The atoms or ions that are stable toward chemical reactivity typically have a completed outer shell or set of eight electrons in the outermost energy level. Examples of this type of structure are the elements in Group VIIIA, Ne, Ar, Kr, Xe, the ions F^{1-}, Cl^{1-}, and O^{2-}. The cations form an octet that matches the completed set of 8 electrons that appears in the noble gas of the previous row, sodium, Na^{1+} has the same electron configuration as Ne in the previous row.

41.

Atomic number	Element name	Number of valence electrons	Period	Metal, M or Nonmetal, NM
6	carbon, C	4	2	NM
12	magnesium, Mg	2	3	M
17	chlorine, Cl	7	3	NM
37	rubidium, Rb	1	5	M
42	molybdenum, Mb	6	5	M
54	xenon, Xe	8 or 0	5	NM

43. Atomic radius is the distance from the center of the nucleus to the outer surface of the electron cloud around the nucleus. Cesium has a larger radius than lithium. The Cs atom has more electrons than Li, it has more energy levels used, and the outer or last electron in Cs is in a higher energy level. There are 55 electrons in cesium and there are only 3 electrons in lithium. Electrons all have a negative charge. Like charged particles repel one another, so the more inner electrons around an atom the greater the repulsions between them and the larger the electron cloud around nucleus. The volume of the atoms increase down the group from Li to Cs.

45. Atomic radius is the distance from the center of the nucleus to the outer surface of the electron cloud around the nucleus. The smaller the radius of the nonmetal atom, the more reactive it is. This is reasonable because the nonmetals are short of an octet. The atomic nucleus will have a stronger attraction for an electron from another atom because there are fewer electrons surrounding the nucleus.

47. There is a stepwise change in proton number across a period. The attraction between the nucleus and the electrons in the atom gradually increases because of this change. This is duplicated in each period so trends in properties appear for each period.

49. K (largest), Al, P, S, Cl (smallest)

51. If element 36 is a noble gas, then element 35 is a halogen and element 37 is an alkali metal.

53. Sulfur, S, and tellurium, Te, are above and below selenium, Se, in the same group.

© 1999 Harcourt Brace & Co. All rights reserved.

Speaking of Chemistry
Structure of the Atom

Name _____

Across
4 The electron _____ for lithium 2-1.
7 Abbreviation for atomic mass unit.
8 Like charges _____.
11 Proust proposed the law of _____ composition in 1799.
12 Scanning tunneling microscope.
15 Part of atom where mass is concentrated.
17 Particles with +1 charge.
19 John _____ proposed an atomic theory in 1803.
20 Oppositely charged particles_____.
21 Person who proposed nuclear model for the atom.

Down
1 The _____ is determined by the number of outer electrons.
2 Person who proposed a model for the energy states of the electron in hydrogen.
3 _____ charged particles repel
4 Electrode that attracts positive charges.
5 Particles with mass of 1.67×10^{-24} g and no charge.
6 _____ rays have no detectable mass or charge.
9 The electrode that attracts negative particles.
10 Atomic force microscope.
13 Amounts of energy needed for quantum jumps.
14 _____ number indicates the number of protons in a nucleus.
16 Particles identical with electrons.
18 Lowest energy state for an atom.

© 1999 Harcourt Brace & Co. All rights reserved.

Bridging the Gap   I   Name _____
Predicting Properties of Elements

The strength of any theory lies in its usefulness to make predictions about new and untried situations. This exercise is aimed at demonstrating the power of the periodic law. The text points out on page 82 that the periodic table is being extended by synthesizing new elements using high energy particles in nuclear accelerators. Element 118 is projected to be a noble gas. This exercise is intended to use the periodic law and trends in the table to predict the properties for a different element, number 119. Information given in the chapter should allow you to predict values and properties for element 119. Write your answers on this sheet.

1. Prediction of Group and Lewis symbol
   What group of representative elements would include atomic number 119 ? Write the Lewis symbol for element 119.(Use Uk for the element symbol.) Explain your reasoning for picking this group. Pick a name and symbol for this element. Tell why you made this choice of a name and symbol.

2. Prediction of atomic properties: approximate atomic weight and atomic radius
   What is the likely atomic weight for atomic number 119? Explain how you made your prediction. Hint: Look at the other elements in group containing Uk-119 and project the trend for ratio of neutrons to protons for successive elements in the group.

   What is the likely atomic radius? Explain how you made your prediction. Check the hint given above.

© 1999 Harcourt Brace & Company. All rights reserved.

3. Prediction of physical properties
   Give predicted or estimated values for the following and describe how you made your choices.

predicted density	predicted melting point

4. Prediction of chemical properties
   What classification would you assign to this element: metal, nonmetal or metalloid? Explain your reasoning.

   Do you predict that element 119 will react with water? Justify your answer.

   What formula do expect for the combination of chlorine and element 119? Write the equation for the possible reaction between chlorine, Cl_2, and element 119.

Bridging the Gap   II
Seeing the Light: Incandescent, Mercury Vapor or Sodium Vapor.

Street lighting and lighting of parking lots is a common public service. The type of electric street lamps used has changed over the years. Incandescent lamps that depended on glowing white hot filaments for light were replaced in the 1950s. Currently, street lamps are mercury vapor lamps, sodium vapor lamps or metal halide lamps (these contain thallium, indium, mercury and sodium). These are cheaper to operate than incandescent lamps because they use less electricity to produce the desired light intensity (lumens per watt). The colors emitted by these lamps are very different. The mercury vapor lamp produces a bluish light. A sodium lamp produces a yellowish color and the metal halide lamps are almost white. The energy carried by a photon of blue light is different from the energy carried by a photon of yellow light. The energy needed to produce these two types of light is also different. Check the text Figure 3.7 on page 51 which shows the continuous spectrum of light and determine which of these two colors carries more energy per photon.

This exercise has four parts. One part is to decide which type of lamp requires less energy. Part two is to determine which kind of street lighting is used in your community by observing the colors of the street lights. Are your street lights mercury vapor lamps, sodium vapor lamps, incandescent lamps, metal halide lamps or are they an assortment of all kinds? A third part is to write a brief argument for standardized type of street lamp. Lastly you are to write one question you would want answered by an expert before you would make a final choice of street lamp to use throughout your community.

© 1999 Harcourt Brace & Company. All rights reserved.

Bridging the Gap   II   Name _____
Seeing the Light:
Incandescent, Mercury Vapor or Sodium Vapor.

1. Type of lamp that uses less energy (include a supporting reason for your answer)

2. Type of street lamp that is most common in your community
   (Tell how you came to this conclusion.)

3. Argument for having a single type of street lamp

4. The question you want an expert to answer so you can make a better choice.

Bridging the Gap   III   Name_____
Internet access to the American Institute of Physics, AIP
A Look Inside the Atom with J. J. Thomson

A Look Inside the Atom
One hundred years ago J. J. Thomson was one of the leading investigators of the structure of the atom. He was Rutherford's teacher at Cambridge University in England. The world had accepted Dalton's atomic theory of indivisible atoms, and people thought that atoms were the smallest particles in the universe. J. J. Thomson's work with cathode ray tubes was the first step into the world of subatomic particles. He investigated the "cathode rays" that were observed when an electric current was passed through a sealed glass tube with a metal electrode at each end. The information needed to complete the following activities is found at the sites whose uniform resource locators (URL) are listed below.

Thomson's speculation
   http://www.aip.org/history/electron/jjhome.htm#cathtube

1.   Thomson speculated that the cathode rays were made of "corpuscles". Where did he believe the matter in the cathode rays originated?

3 Experiments, 1 Big Idea
   http://www.aip.org/history/electron/jj1897.htm

2.   In the years 1895-1897, J.J. Thomson did a series of three experiments on the nature of cathode rays. When Thomson did his first experiment he used an "electrometer" to measure electric charge when the cathode ray beam was allowed to strike the electrometer and then again when a magnet was used to turn the beam away from the electrometer. What did this experiment show?

3. What "trick" did Thomson use in his second experiment to successfully bend a cathode ray beam with an electric field? What was his conclusion about the matter in the cathode ray beam after his second experiment?

4. What information about the cathode rays did Thomson learn in his third experiment?

The Legacy for Today
http://www.aip.org/history/electron/jjlegacy.htm

5. What modern devices and technologies rely on Thomson's 1897 experiments and the information he obtained?

4 Nuclear Changes

4.1 The Discovery of Radioactivity

Becquerel's discovery of radioactivity in 1896 is described. Studies by Rutherford and Villard on the penetrating power of alpha, beta and gamma rays are summarized. Alphas can be stopped by paper, betas can be stopped by 2.5 cm to 8.5 cm of air and gammas can pass through more than 22 cm of steel.

Objectives
After studying this section a student should be able to:
- describe Becquerel's experiment that proved that phosphorescence and radioactivity are different phenomena
- describe the mass and charge of alpha and beta particles
- tell how the penetrating powers of alpha, beta and gamma rays differ
- explain what is meant by the term ionizing radiation

Key terms
X rays	phosphorescence	emitted radiation
alpha, beta, gamma rays	penetrating ability	ionizing radiation

4.2 Nuclear Reactions

The differences between chemical reactions and nuclear reactions are summarized. The method for balancing nuclear reactions is described and illustrated for the alpha and beta emissions. The rules for balancing decay reactions are illustrated. The total of mass numbers for reactants must equal the total of the mass numbers for products. The total of atomic numbers for reactants must equal the total of atomic numbers for products.

Beta emission: the mass number stays the same and atomic number increases by one

$$^{210}_{82}Pb \rightarrow {}^{210}_{83}Bi + {}^{0}_{-1}e$$

Alpha emission: the mass number decreases by 4 and the atomic number decreases by 2

$$^{218}_{84}Po \rightarrow {}^{214}_{82}Pb + {}^{4}_{2}He$$

Beta particle emission is explained as the break up of a neutron. $^{1}_{0}n \rightarrow {}^{0}_{-1}e + {}^{1}_{1}H$

Objectives
After studying this section a student should be able to:
- state the definition for radioactivity, nucleons and transmutation
- describe what a nuclear reaction is and give an example
- tell how a nuclear reaction differs from a chemical reaction
- complete a nuclear equation given the starting nucleus and type of emission
- identify a balanced nuclear equation
- tell how nuclear charge and mass change when either an alpha, $^{4}_{2}He$, or beta particle, $^{0}_{-1}e$, is emitted

Key terms
atomic number	transmutation	nucleons
mass number	nuclear equation	beta-emitter
alpha-emitter	$^{4}_{2}He$	$^{0}_{-1}e$

4.3 The Stability of Atomic Nuclei

The stability of nucleons is explained in terms of the plot of number of neutrons versus number of protons for known isotopes. Isotopes are defined as atoms like $^{2}_{1}H$ and $^{3}_{1}H$ that have the same atomic number but different mass numbers. Stability is explained in terms of the position of an isotope relative to the band of stability. Nuclei with too many neutrons emit betas. All elements beyond atomic number 83 are radioactive, and most decay by alpha emission. Nuclei with too few neutrons emit a positron. Positron emission reaction is illustrated below.

$$^{30}_{15}P \rightarrow {}^{30}_{14}Si + {}^{0}_{+1}e$$

Objectives
After studying this section a student should be able to:
- give the standard symbol for atomic number
- describe the properties of a positron
- state the definition of isotopes and give an example like $^{3}_{1}H$ and $^{2}_{1}H$
- tell how the atomic number and the mass number change for an isotope after an event such as alpha emission, beta emission, positron emission
- tell how neutron number and atomic number are related in stable nuclei
- predict the kind of particle emitted by an atom with too few neutrons i.e. $^{13}_{7}N$
- predict the kind of particle emitted by an atom with too many neutrons i.e. $^{210}_{82}Pb$
- tell what type of emission is expected for atoms that are too big such as $^{230}_{90}Th$

Key terms
stability of nuclei   alpha emission   positron emission
isotopes   positron or $^{0}_{+1}e$ beta emission

4.4 Activity and Rates of Nuclear Disintegrations

Activity measures the number of nuclear disintegrations per unit time. The activity unit, the curie is defined as 37 billion disintegrations per second(dps). Cobalt-60 decay and the emission of gamma rays are shown. Examples show how radioactive source activity and mass change with time. Half life is defined. Graphs of activity versus half life are explained. The natural decay series are discussed and the specific steps in the uranium-238 decay series are graphed.

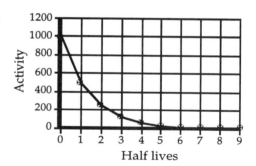

Objectives
After studying this section a student should be able to:
- give definitions for half life and activity
- tell what a curie (Ci) and a microcurie (µCi) equal in terms of disintegrations per second
- predict the activity of a source if given initial activity, elapsed time and the half life period
- graph the mass of radioactive material versus time if given initial mass and half life period
- tell what is meant by a natural radioactive decay series

Key terms
activity   half-life   curie, Ci
microcurie, µCi, 37,000 dpi cobalt-60   gamma emitter
decay series   uranium series

4.5 Artificial Nuclear Reactions

Irene Curie Joliot and Frederic Joliot produced the first radioisotope to be made artificially by bombarding Al nuclei with high energy alpha particles. Currently, artificial transmutation has produced over 1000 new radioactive isotopes.

Objectives
After studying this section a student should be able to:
- describe how Rutherford produced the first artificial nuclear change
- give the definition for transmutation and describe how transmutation is produced
- describe the roles of Irene Curie Joliot and Frederic Joliot in making artificial radioisotopes

Key terms and equations
artificial nuclear reaction   artificial transmutation   bombardment reaction

4.6 Transuranium Elements

The transuranium elements, those elements with Z greater than 92, are all artificial elements. The production of the highly-toxic isotope plutonium-239 from uranium-238 is illustrated. The bombardment reactions used to make elements 101 to 113 are discussed. The controversy over naming new elements is described.

Objectives
After studying this section a student should be able to:
- give the definition of a transuranium element
- identify a transuranium element if given its symbol or atomic number
- tell what kinds of projectile particles are used to produce elements up to Z = 101
- balance a nuclear bombardment equation and identify the product nucleus given the target and bombardment particle

Key terms and equations
transuranium elements   bombardment experiment   plutonium

4.7 Radon and Other Sources of Background Radiation

This section describes background radiation, from both man-made and natural sources. The normal background radiation level is approximately 2 or 3 dps. A detailed discussion identifies radon as the dominant natural source of ionizing radiation. The half life of radon-222 and the hazards posed by it and its daughters are described. The cancer risk posed by radon-222 is explained. The U.S. Environmental Protection Agency (EPA) defines an action level for mitigating radon levels as 4 pCi/L. (pico = 1×10^{-12})

Objectives
After studying this section a student should be able to:
- name an example of a natural and a man-made radiation source
- describe the radon decay reaction and explain why it is a lung cancer hazard
- explain how background radiation arises
- tell which regions of the United States have the greatest radon potential

Key terms and equations
ionizing radiation   background radiation   picocurie, pCi
epithelial cells   EPA   radon, $^{222}_{86}Rn$

4.8 Useful Applications of Radioactivity

This section reviews some beneficial uses of radioactivity. The use of cobalt-60 and cesium-137 gamma irradiation in food preservation is discussed. Technetium-99m in medical imaging of bone abnormalities is explained.

© 1999 Harcourt Brace & Company. All rights reserved.

Objectives
After studying this section a student should be able to:
- describe how foods can be preserved using gamma rays from sources like cobalt-60
- give a definition of "radiolytic products"
- name a diagnostic radioisotope and tell what organ it is used to image
- give examples of foods that are preserved using gamma irradiation

Key terms and equations
gamma rays	irradiated food	FDA
radiolytic products	medical imaging	diagnostic radioisotopes

4.9 Energy from Nuclear Reactions
This section defines nuclear fission and fusion. It illustrates and describes thermal neutrons, fission products, critical mass, and chain reactions. An historical summary of the practical use of fission is included. The source of nuclear energy from the conversion of mass to energy is explained. The relative stability of isotopes is diagrammed and explained in terms of binding energy per nuclear particle (nucleon).

Objectives
After studying this section a student should be able to:
- give definitions for nuclear fission, fission products, critical mass, and chain reaction
- describe the fission process and the role of thermal neutrons
- tell why natural uranium must be enriched to support fission
- state the Einstein mass energy equation
- give the definitions for nucleon and binding energy per nucleon
- tell how to calculate the energy yield from a change in nuclear mass
- describe the origin of nuclear binding energy

Key terms
fission	thermal neutrons	fission products
chain reaction	critical mass	subcritical masses
enriched uranium	fission bomb	Einstein's equation
mass difference	$E = mc^2$	UF_6
$^{238}_{92}U$	$^{239}_{94}Pu$	$^{235}_{92}U$
binding energy	binding energy per nucleon	

4.10 Useful Nuclear Energy
This section describes Enrico Fermi's work on controlled nuclear fission. It gives definitions for nuclear reactor components. The layout of a nuclear power plant is diagrammed. Significant nuclear power plant accidents such as Three Mile Island in the U.S. and Chernobyl in the former Soviet Union are described. The hazards of a nuclear reactor core meltdown are explained. Nuclear wastes are defined. Problems associated with nuclear waste storage and disposal are discussed. Nuclear fusion research, controlled fusion and plasmas are described.

Objectives
After studying this section a student should be able to:
- describe the function for each of the following in a nuclear reactor:
    moderator, neutron absorber, shielding, heat-transfer fluid
- describe the scope of the 1979 accident at the Three Mile Island nuclear plant
- describe the scope of the 1986 accident that occurred in the Soviet reactor at Chernobyl
- tell what happens in a nuclear reactor core meltdown
- tell why nuclear power plant wastes are a public concern

describe proposals for the safe disposal of nuclear wastes
describe the nuclear fusion process and tell why it is so attractive
define the terms thermonuclear and plasma
explain why a "magnetic bottle" is needed in plasma and fusion research

Key terms

atomic reactor	shielding	neutron absorber
moderator	core	fuel rod
reactor fuel	heat-transfer fluid	ordinary (light) water
heavy water	core meltdown	unavoidable Pu production
fission products	Three Mile Island	Chernobyl Unit 4 Reactor
$^{238}_{92}U$	$^{235}_{92}U$	$^{239}_{94}Pu$
spent fuel rods	high-level wastes	weapons-grade plutonium
glass logs	nuclear waste products	underground repository
$^{137}_{55}Cs$	$^{131}_{53}I$	Yucca Mountain, Nevada
controlled fusion	deuterium & tritium	magnetic bottle
thermonuclear bomb	plasma	

Additional readings

Anderson, Christopher & Michael Cross. "Fusion Research at the Crossroads" <u>Science</u> April 29, 1994: 648.

Armbruster, Peter & Gottfried Munzenberg. "Creating Superheavy Elements" <u>Scientific American</u> May 1989: 66.

Ghiorso, Albert & Glenn T. Seaborg. "The Newest Synthetic Elements" <u>Scientific American</u> Dec. 1956: 66.

Hamilton, J. H. & J. A. Maruhn. "Exotic Atomic Nuclei" <u>Scientific American</u> July, 1986: 80

Neher, Paul "Radon Monitor." <u>Electronics Now</u> Feb. 1994: 66.

Platzman, Robert L. "What Is Ionizing Radiation?" <u>Scientific American</u> Sept. 1959: 74.

"Radon: Worth Learning about.." (Includes an article on home radon detectors) (Special Report: Safe at Home) <u>Consumer Reports</u> July 1995: 464(2).

Spoerl, Edward. "The Lethal Effects Of Radiation" <u>Scientific American</u> Dec. 1951: 22.

Answers to Odd Numbered Questions for Review and Thought

1. Gamma rays can pass through more than 22 cm of steel and about 2.5 cm of lead. Alpha particles can be stopped by a thin piece of paper or 2 to 8 cm of air. Beta particles have greater penetrating power and can travel through as much as 15 meters of air. Beta particles can penetrate paper but can be stopped by a thin (0.5 cm) sheet of lead. Gamma radiation would pass through the paper and the thin sheet of lead and would require 2.5 cm of lead to stop it.

3. The more hazardous radioisotope is $^{222}_{86}Rn$ with a short half-life of 3 days because more ionizing radiation is emitted over a shorter time period. This means anyone near the source during this time will be exposed to a higher number of emitted particles and a greater risk.

5. Radon-222 is a health hazard because it undergoes alpha decay. It is a gas and can be inhaled. The high density makes it difficult to exhale and the emitted alpha particles have enough energy to penetrate the epithelial lining of the lungs to produce cell damage. Radioactive radon daughters are not gases and will stay in the lungs where they can decay to cause more damage.

7. The law of conservation of matter requires the mass numbers of the products equal the sum of the mass numbers of reactants. The mass number on the phosphorus atom should be 30
$$^4_2He + {}^{27}_{13}Al \rightarrow {}^{30}_{15}P + {}^1_0n$$

9. Seaborg proposed that thorium, Th, was the first element of a different, new series of elements and did not belong to the same group as hafnium, Hf, in Group IVB. He proposed that Th and the transuranium elements belonged as a group under the rare earth elements Ce, etc.

11. The uranium series refers to a sequence of radioactive decay steps that starts with radioactive U-238 and ends with stable Pb-206.

13. Radioisotopes are used for diagnosis and treatment of diseases.

15.
		charge	relative mass
a.	beta particle	-1	0
b.	alpha particle	+2	4
c.	gamma rays	0	0
d.	positrons	+1	0
e.	neutrons	0	1

17. The gamma rays can damage DNA sequences necessary for cell reproduction, thereby slowing or stopping cell reproduction. Cancer cells reproduce faster than normal cells and are more sensitive to this because the DNA sequence is needed for duplication of the genetic code more often in the rapidly multiplying cancer cells.

19. Natural radiation in food, water and air exceeds the exposure from weapons fallout. The radiation coming from food, water, and air equals 24 mrem. The exposure from weapons fallout is 4 mrem per year. The argument against atmospheric nuclear testing holds that the normal background cannot be altered, and the human addition of radiation introduces an additional burden or risk.

21. $^{249}_{98}Ca + {}^{15}_7N \rightarrow {}^{260}_{105}Hn + 4\,{}^1_0n$ Four neutrons are expected because the mass numbers must equal 264 on both sides and the charges must add to 105.

23. a. Becquerel discovered ionizing radiation was emitted by uranium compounds.
    b. Marie Curie demonstrated that radioactivity was an atomic property. She isolated radium and discovered polonium, Po.

c. Rutherford proposed that radioactive decay was a natural change of an isotope of one element into an isotope of another element.

25. The emission of a beta particle has no effect on the mass number of the daughter because the mass number for a beta particle is zero.
    $$^{211}_{82}Pb \rightarrow {}^{211}_{83}Bi + {}^{0}_{-1}e$$

27. Nuclear energy from fission has risks of a catastrophic accident, radioactivity releases from plutonium production, problems of disposal of radioactive wastes. The plutonium can be used for construction of nuclear weapons and could increase the possibility of more nations having nuclear weapons. Some people worry that terrorists might steal the plutonium to make bombs.

29. a. The "reprocessing of nuclear fuels" refers to the separation of useful plutonium and fissionable uranium from other nuclides in spent fuel elements.
    b. A number of dangers are linked to the reprocessing. The plutonium may be diverted for nuclear weapons. There maybe a breach of the process and radioactive material may escape into the environment. Plutonium itself is extremely toxic to humans.

31. Fission and fusion both rely on conversion of mass to energy. Fission requires splitting heavy nuclei into smaller ones while fusion involves joining small nuclei together to produce a larger nucleus.

	Fuels	Benefits	Problems	Current Status
Fission	Uranium Plutonium	renewable from breeder reactors	waste storage and disposal Pu toxicity	commercially available, proven method
Fusion	Deuterium Tritium	unlimited fuel supply	radioactive hardware from power plants, no containment vessel available	research only

33. a. $^{64}_{29}Cu \longrightarrow {}^{64}_{30}Zn + {}^{0}_{-1}e$ beta emission

    b. $^{69}_{30}Zn \longrightarrow {}^{69}_{31}Ga + {}^{0}_{+1}e$ positron emission

    c. $^{131}_{53}I \longrightarrow {}^{131}_{54}Xe + {}^{0}_{-1}e$ beta emission

35. a. The half-life of $^{239}_{94}Pu$ is 24,000 years.
    b. Plutonium is a problem because it does not dissipate quickly. The plutonium is chemically extremely toxic.

37. The "magnetic bottle" uses magnetic fields produced by strong electromagnets to contain a plasma by interacting with the magnetic fields generated by the ions in the plasma. This keeps the high temperature plasma from physical touching a container and dissipating the energy stored in the plasma.

© 1999 Harcourt Brace & Company. All rights reserved.

Answers for Problems

1. The number of atoms will decline by 1/2 after each half-life. The 300,000 atoms will decrease to 150000 atoms after one half-life; 75000 atoms after two half-lives; 37500 atoms after three and 18750 atoms after four. The graph illustrates the decrease in the number of atoms.

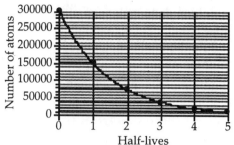

   Alternatively, the number of atoms remaining after "n" half-lives can be calculated from the equation: amount remaining = (initial amount)$\left(\frac{1}{2}\right)^n$

   amount remaining = $(300,000 \text{ atoms})\left(\frac{1}{2}\right)^4 =$
   $(300,000 \text{ atoms})\left(\frac{1}{2}\right)\left(\frac{1}{2}\right)\left(\frac{1}{2}\right)\left(\frac{1}{2}\right) = (300,000 \text{ atoms})\left(\frac{1}{16}\right) = 18750$ atoms

2. If initially there were 200,000 atoms of $^{238}_{92}U$, there will be only 100,000 atoms left after one half-life of 4.5 billion years. There would be 100,000 alpha particles produced from the decay of $^{238}_{92}U$ during this time. The decay equation shows a one to one ratio between uranium's lost and alphas produced. $^{238}_{92}U \rightarrow {}^{234}_{90}Th + {}^{4}_{2}\alpha$

   A graph can be made showing the number of particles initially and after three half-lives.

   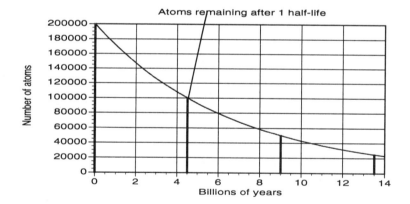

   amount remaining = $(200,000 \text{ atoms initially})\left(\frac{1}{2}\right)^1 = (100,000 \text{ atoms remaining})$

Speaking of Chemistry

Nuclear Changes

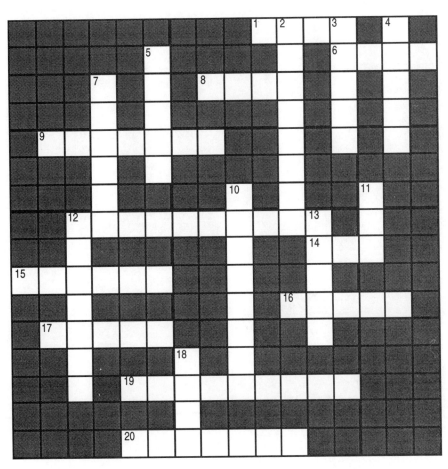

Across
1. High energy electron
6. Element 82 at end of uranium series
8. One of discoverers of fission in 1938.
9. Element with atomic number 92
12. Radiation naturally present at 2-3 dps
14. Environmental Protection Agency
15. Radon-222 can cause lung _____
16. Asian country that produces 33% of its electricity using nuclear power
17. Beta particles have _____ charge
19. Site of 1986 Ukrainian reactor accident
20. Atoms like ^3_1H and ^2_1H

Down
2. Proposed $E = mc^2$ theory
3. ^4_2He particle
4. Radioactive gas present in rock, Z = 86
5. Received Nobel prize and isolated radium
7. A state of matter containing nuclei and unbound electrons
10. Particle with mass of electron but +1 charge
11. Food and Drug Administration
12. Nuclear _____ energy holds nucleus together
13. Radioactive _____ is process in which an isotope of one element changes into an isotope of another
18. Charge on a neutron

© 1999 Harcourt Brace & Company. All rights reserved.

Bridging the Gap I Name _____
Your Annual Dose and the U.S. Annual Average

Every day you are exposed to radiation from radioactive materials in your environment. These radioactive materials are found in soil, rocks, building materials, food, air, and water. Cosmic rays, medical tests, TV, and airplane travel all influence the total amount or dose you receive each year. You can estimate your annual dose of radiation using the information below. This table is adapted from A.R. Hinrichs: Energy, pp. 335-336. Philadelphia, Saunders College Publishing, 1992.

Estimate of your annual dose

	Sources of Radiation	Annual dose in mrem
Where You Live	Location: Cosmic radiation at sea level	26
	For your elevation (in feet) add this number of mrem Elevation mrem Elevation mrem Elevation mrem 1000 2 4000 15 7000 40 2000 5 5000 21 8000 53 3000 9 6000 29 9000 70	_____
	Ground: U.S. average	26
	House construction: For stone, concrete, or masonry building add 7	_____
What You Eat, Drink, and Breathe	Food, water air: U.S. average	24
	Weapons test fallout	4
How You Live	X ray and radiopharmaceutical diagnosis: Number of chest x rays _____ x 10 Number of lower gastrointestinal tract X rays _____ x 500 Number of radiopharmaceutical examinations _____ x 300 (Average dose to total U.S. population = 92 mrem)	_____ _____ _____
	Jet plane travel: For each 2500 miles add 1 mrem	_____
	TV viewing: Number of hours per day ____ x 0.15	_____
How Close You Live to a Nuclear Plant	At site boundary: average number of hours per day _____ x 0.2 One mile away: average number of hours per day _____ x 0.02 Five miles away: average number of hours per day _____ x 0.002 Note: Maximum allowable dose determined by "as low as achievable"(ALARA) criteria established by the U.S. Nuclear Regulatory Commission. Experience shows that your actual dose is substantially less than these limits.	_____ _____ _____
	Your total annual dose in mrem	_____

© 1999 Harcourt Brace & Company. All rights reserved.

Comparing your estimated dose with the U.S. Annual Average

The average annual dose in the U.S. is 180 to 200 mrem. How does your annual dose compare with this average? Is your dose high or low?

Principle source of exposure

What source is the largest contributor to your annual dose?

How exposure changes for a frequent flyer

Many executives and salespersons do a lot of flying as part of their jobs. How many additional mrem would a frequent flyer accumulate if this person took the trips listed below in one year?

Trip	Number of round trips	One way mileage
Chicago to San Francisco	2	2108
Chicago to Miami	2	1338
Chicago to Los Angeles	4	1990
Chicago to Seattle	2	2043
Chicago to New York	4	794
Chicago to New Orleans	2	925
Chicago to Denver	1	1037
Chicago to Atlanta	2	695
Chicago to Phoenix	3	1776

What percentage would this be of the average annual dose of 180 mrem?

Bridging the Gap   II   Name_____
Internet access to the Environmental Protection Agency, EPA
A Citizens Guide to Radon
Internet access to the University of Maine: Radon Information

If you do not have access to the internet you can get information by calling the toll-free National Radon Hotline: 1 800/SOS-RADON

The EPA
The Environmental Protection Agency is readily accessible using the internet.
The general URL for the EPA is
> http://www.epa.gov/

Information about radon is available at the following URL.
> http://www.epa.gov/docs/RadonPubs/citguide.html

The content of any displayed web page can be down loaded to disk as text or it can be printed. The underlined subject headings on EPA homepage are interactive and lead to additional sites. The interactive topics can be selected by clicking on the underlined text. Each one of these links to another page with a different URL.

Remember, webpages are constantly updated. The appearance and content of a site will change when the site is updated.

This EPA site at
> http://www.epa.gov/docs/RadonPubs/physic.html#Whatis

or this excellent site maintained by the University of Maine
> http://www.physics.csbsju.edu/MNradon/maps/epa.html

can be used to answer the questions in this activity.

Remember, if you do not have access to the internet, you can get information by calling the toll-free EPA National Radon Hotline: 1 800/SOS-RADON

1. Use information at this URL to answer the following question.
   > http://www.epa.gov/docs/RadonPubs/citguide.html

   What does the EPA say about the number of deaths per year that are the result of radon exposure?

EPA and the University of Maine

Use either the Maine website or the additional EPA site to answer the exercise questions.
  University of Maine
      http://www.physics.csbsju.edu/MNradon/maps/epa.html

  EPA site titled "A Physician's Guide -Radon"
      http://www.epa.gov/docs/RadonPubs/physic.html#Whatis

2. Click on the link • What is Radon?
   Identify the particle emitted by radon as it decays?

3. Click on the link • How Does Radon Get Into Homes?
   Name the three ways radon can enter a home.

4. Click on the link •Projected Health Risks
   What is the health risk posed by breathing air contaminated by radon?

5. Give the URL for a site related to radon that you found interesting.

      http://_____

   What makes it interesting to you? Would you recommend it to someone else? You can send your opinion to the Webmaster of the site. They appreciate comments.

5   Chemical Bonding and States of Matter

5.1   Ionic Bonds

In this section ionic bonds, and salts are defined; the octet rule is applied in the formation of Na^+ and Cl^- ions from the atoms. The section explains how positive ions are formed when electrons are "peeled" off of a neutral atom. The sodium atom can be converted to a Na^+ ion in this way. Here the one valence electron is removed. Negative ions are formed when a neutral atom gains one or more electron. Chlorine has seven valence electrons. It can accept one more electron to complete the octet to make a Cl^- ion. Ionic bonds are formed when positive and negative ions are allowed to interact. The formula unit for ionic substances is a neutral combination of positive and negative ions where the total positive charge is canceled by the total negative charge. Ionic compounds are defined, and the concept of a formula unit is explained.

Objectives
After studying this section a student should be able to:
    state the definition of a salt
    give definitions for ionic bond and ionic compound
    tell what forces hold an ionic crystal together
    explain why ion charges differ from atomic charges

Key terms
salts	valence electron	metal	nonmetal
octet rule	ionic compound	ionic bond	electrostatic attraction
formula unit			

5.2   Ionic Compounds

This section relates the octet rule to the Lewis dot symbols for positive and negative ions, shows how ions combine to form neutral compounds, and gives general rules for the formation of ionic compounds and for finding their formulas. It gives the rules for naming binary ionic compounds. Finally, it shows how polyatomic ions combine to form neutral formula units.

Objectives
After studying this section a student should be able to:
    write the Lewis dot symbol for a neutral atom
    state the octet rule
    decide how many electrons an atom will lose or gain to form an octet and predict the
        charge for the ion formed when these electrons are lost or gained
    write the Lewis dot symbol for the anion or the cation formed from a particular atom
    predict the formula for the binary compound formed by a given pair of elements
    give the formula and charge for common ions in the A groups
    give the definition for a binary compound
    explain how anion names are formed from element names
    state the definition of polyatomic ion
    predict the formula for a combination of a given metal cation and polyatomic anion
    name a binary compound from its formula
    write the formula for a binary compound from its name
    name the common polyatomic anions like those in Table 5.1

© 1999 Harcourt Brace & Company. All rights reserved.

Key terms

Lewis dot symbol	noble gas configuration	A group nonmetal
anion	negative ion	"ide" ending
A group metal	cation	Roman numeral and charge
binary compound	polyatomic ion	hydroxide ion
carbonate ion	hydrogen carbonate	acetate ion
phosphate ion	nitrate ion	nitrite ion
sulfate ion	sulfite ion	dihydrogen phosphate ion

5.3 Covalent Bonds

This section gives a definition for the covalent bond, describes single covalent bonds, defines unshared (lone pair, nonbonding) electrons, and shows how to predict molecule formulas for nonmetal-nonmetal combinations like $:\!\overset{\cdot}{\underset{\cdot}{N}}\!\cdot$ and $H\cdot$ to make $H\!:\!\overset{H}{\underset{\cdot\cdot}{\overset{\cdot\cdot}{N}}}\!:\!H$ It describes saturated hydrocarbons as compounds with only single covalent bonds, gives the formula C_nH_{2n+2} for saturated hydrocarbons, describes multiple covalent bonds (double and triple bonds), summarizes properties of alkenes and alkynes and defines unsaturated hydrocarbons. The section describes how to name binary molecular compounds, and gives definitions for the prefixes used in naming. It shows how to relate single, double and triple bonds to Lewis dot structural formulas for molecules.

Objectives

After studying this section a student should be able to:
- state the definition for a covalent bond
- draw the Lewis symbol for a molecule if given the formula such as H_2O
- describe the difference between a nonbonding pair of electrons and a bonding pair
- give definitions for hydrocarbons, alkanes, and saturated hydrocarbons
- describe single, double, and triple covalent bonds
- give a definition for bond energy and explain what the values represent
- describe what unsaturated means
- identify the types of bonds in a molecule if given the structural formula
- match prefixes and number of atoms i.e.. mono- one; di- two; tri- three
- name representative binary molecular compounds if given the formula

Key terms

Lewis structures	covalent bond	nonbonding pair
hydrocarbon	single covalent bond	saturated hydrocarbon
alkane, C_nH_{2n+2}	bond length	unsaturated hydrocarbon
multiple covalent bond	double bond	triple bond
alkene, C_nH_{2n}	prefixes	mono- for one
binary molecular compounds	tri- for three	tetra- for four
di- for two	hexa- for six	hepta- for seven
penta- for five	nona- for nine	octa- for eight

5.4 Shapes of Molecules

This section describes a model for predicting shapes of molecules and ions, and reviews definitions of bonding and nonbonding pairs of electrons. It introduces bond angle measurements and describes the linear, bent, triangular planar, triangular pyramidal and tetrahedral shapes.

Objectives
After studying this section a student should be able to:
- predict the shape of a molecule or ion using its Lewis structure
- state the difference between the terms "electron-pair geometry" and "molecular geometry"
- use the electron-pair geometry to predict bond angles

Key terms and equations
linear geometry
bent geometry
tetrahedral geometry
molecular geometry

triangular planar geometry
triangular pyramidal geometry
electron pair geometry

5.5 Polar and Nonpolar Bonding

Bond polarity is explained for both nonpolar and polar covalent bonds. The convention using ↦ to indicate bond polarity is introduced. Molecular shapes are related to bond polarity to explain the polarity of molecules. Carbon dioxide is used to show how the symmetry of a molecule can cancel out the effects of polar bonds. Water is used to show how a polar molecule results from bent structures and polar bonds.

— slightly negative end of molecule
— slightly positive end of molecule

Objectives
After studying this section a student should be able to:
- tell what atom in a compound attracts electrons most strongly given the formula
- describe how the attraction for electrons changes for elements in the same group
- describe the trend for the attraction for electrons across the periodic table
- give the basis for deciding on the polarity of a bond
- show the polarity of a bond for a given pair of atoms using the ↦ symbolism
- describe the difference between a polar and a nonpolar bond
- tell if a molecule is polar or nonpolar using molecule shape
- tell what "equal sharing" means in covalent bonding

Key terms and equations
equal sharing
polar molecule
symmetrical

nonpolar covalent bond
unequal sharing
electron shift

polar covalent bond
nonpolar molecule
partial charges

5.6 Properties of Molecular and Ionic Compounds Compared

This section tells how the strong electrostatic forces in ionic solids make them brittle and hard. The weaker attractions between polar and nonpolar molecules lead to softer, lower melting substances. Electrical conductivity of solid ionic compounds, molten ionic compounds, and solutions are linked to the concept of electrolytes and nonelectrolytes.

© 1999 Harcourt Brace & Company. All rights reserved.

Objectives
After studying this section a student should be able to:
- outline the properties of ionic and molecular compounds in terms of hardness, melting point range, boiling point range, solubility in water, and conductivity of the substance when molten (melted)
- tell how an electrolyte behaves
- describe how a nonelectrolyte behaves

Key terms and equations
molecular compound   ionic compound   brittle
conductivity   electrolyte   nonelectrolyte

5.7 Intermolecular Forces
This section gives definitions for intermolecular forces. The attractions between polar molecules are described. The origins of attractive forces between nonpolar molecules are discussed. Hydrogen bonding between polar molecules like HF, NH_3 and H_2O is described in detail. Hydrogen bonding only occurs in molecules where H is bonded to F, O or N.

Objectives
After studying this section a student should be able to:
- give a definition for intermolecular forces
- give a rough idea of the strength of intermolecular forces compared to covalent bonds
- describe how polar molecules like SO_2 interact using words and sketches
- describe the origin of intermolecular forces between nonpolar molecules
- tell how molecular size and valence electrons affect forces between nonpolar molecules
- give a definition for hydrogen bonding and tell how it affects the properties of H_2O
- state the requirements for hydrogen bonding

Key terms
intermolecular forces   attractive forces   condensation
polar molecule   partially negative region   partially positive region
nonpolar molecule   induced attractive forces   hydrogen bonding

5.8 The States of Matter
The constant motion of molecules in solids, liquids, and gases is described. The distances traveled by particles in solids, liquids, and gases are compared. The relationship between temperature and movement is explained. Freedom of movement and temperature are related to condensation of gases and the melting of solids.

Objectives
After studying this section a student should be able to:
- tell how the freedom of movement of particles differs for solids, liquids, and gases
- describe how increasing temperature influences particle velocity
- explain why intermolecular forces result in different condensation temperatures for gases
- explain why liquids are free to flow

Key terms
states of matter   gas   solid
liquid   fluid   regular arrangement

5.9 Gases and How We Use Them

The effect of temperature on gas pressure is explained. Gas compressibility and Boyle's Law are described in terms of kinetic molecular theory. The large distances between gas molecules is used to explain the miscibility of gases. Gas diffusion and liquid volatility are related to the distribution of odors.

Objectives
After studying this section a student should be able to:
- describe how frequency of gas particle collisions relates to temperature
- explain why, at constant pressure, gas volume decreases with a decrease in temperature
- explain why gases are miscible
- state the definition for the term "volatile"
- tell what gaseous diffusion means

Key terms

pressure	compressibility	Boyle's law
volatile	gaseous diffusion	miscibility

5.10 Water

This section describes the physical properties of water and relates them to the unusually strong intermolecular (hydrogen bonding) attractions between water molecules. The following properties are reviewed: normal boiling point, vapor pressure, density, heat capacity, heat of vaporization, surface tension, and ability to act as a solvent.

Objectives
After studying this section a student should be able to:
- explain why ice (solid water) has a lower density than liquid water
- give definitions for heat capacity, heat of vaporization and surface tension
- give a definition for vapor pressure
- tell why the normal boiling point for a liquid indicates volatility
- examine a graph of vapor pressure versus temperature for a series of liquids and identify the most volatile
- describe how vapor pressure of a liquid depends on temperature
- define normal boiling point and boiling point and tell how these differ

Key terms

heat capacity	density	heat of vaporization
water vapor	vapor pressure	boiling point
hydrogen bonding	surface tension	normal boiling point

5.11 Solutions

This section gives definitions for terms that describe solutions and the dissolving process. The effects of structure on solubility are illustrated using the decreasing water solubility of alcohols with increased carbon count. The link between increased gas pressure and greater gas solubility is discussed. The decease in gas solubility in water with increased temperature is explained.

Objectives
After studying this section a student should be able to:
- give definitions for solvent, solute, solubility, solution, insoluble
- state definitions for saturated and unsaturated solutions
- describe how solubility of alcohols in water depends on molecule size
- explain why small alcohol molecules are more water soluble than large ones
- describe how gas solubility in water depends on temperature
- tell how gas pressure influences gas solubility and give an example

© 1999 Harcourt Brace & Company. All rights reserved.

Key terms

solution	homogeneous mixture	solvent
solute	solubility	miscibility
saturated	unsaturated	insoluble
gas solubility	gas solubility & pressure	gas solubility & temperature

5.12 Solids

This section includes a comparison of the freedom of movement of particles in solids, liquids, and gases. The differences in the hardness of graphite and of diamond are described and explained. The regular arrangement of particles in solids is outlined. Melting, crystallization, and sublimation are described. The sublimation of CO_2 and the principles behind frost-free refrigerators are discussed. The connection between melting points and intermolecular forces are explained.

Objectives

After studying this section a student should be able to:
- describe the structure for diamond and explain why diamond is very hard
- describe the bonding and structure for graphite and explain why it is soft
- define melting point
- describe crystallization
- describe how sublimation of H_2O plays a role in frost free refrigerators

Key terms

random motion	vibration	rotation
molecular motion	crystals	crystallization
sublimation	melting point	

Additional readings

Amoore, John E.., et al. "The Stereochemical Theory Of Odor" <u>Scientific American</u> Feb. 1964: 42.

Breslow, Ronald. "The Nature Of Aromatic Molecules" <u>Scientific American</u> Aug 1972: 32.

Colson, Steven D. and Thom H. Dunning Jr. "The Structure of Nature's Solvent: Water." <u>Science</u> July 1, 1994 : 43.

Curl, Robert F. and Richard E. Smalley. "Fullerenes" <u>Scientific American</u> Oct. 1991: p 54.

Derjaguin, Boris V. "The Force Between Molecules" <u>Scientific American</u> July, 1960: p 47.

Lambert, Joseph B. "The Shapes Of Organic Molecules" <u>Scientific American</u> Jan 1970: p 58.

Zewail, Ahmed H. "The Birth Of Molecules" <u>Scientific American</u> Dec. 1990: p 76.

Pool, Robert. "Atom Smith. (Dick Siegel Constructs Materials One Molecule at a Time)" <u>Discover</u> Dec. 1995: 54.

Answers to Odd Numbered Questions for Review and Thought

1.
   a. A cation is an ion with a positive charge, like Ca^{2+}.
   b. An anion is an ion with a negative charge, like Cl^{1-} or CO_3^{2-}.
   c. Atoms react to acquire an electron configuration with 8 electrons in the outermost shell to match the configuration of the nearest noble gas. Sodium 2-8-1 will lose one electron to match the neon configuration 2-8. Chlorine 2-8-7 will gain one electron to form match argon, 2-8-8.. The valence electron shell
   d. The formula unit is the simplest element ratio for an ionic compound like NaCl instead of Na_2Cl_2.

3.
   a. A shared pair is a pair of electrons shared between two atoms like in H:Cl

   shared pair — H:Cl: ← unshared pairs

   b. Four electrons (2 pairs) shared by two atoms as between carbons in $H_2C::CH_2$
   c. Six electrons (3 pairs) shared between two atoms as in HC:::CH
   d. An unshared pair is a pair of valence electrons on an atom that are not shared with another atom. See 3a above
   e. A single bond is one pair of electrons shared between two atoms as in H:H
   f. A multiple bond is either a double or a triple bond.

5.
   a. A nonpolar bond exists between two atoms that share electrons equally.
   b. A polar bond exists between two atoms that do not attract shared electrons equally.

7.
   a. A hydrocarbon is a compound consisting only of carbon and hydrogen like ethane, CH_3CH_3; ethylene, CH_2CH_2 ; acetylene, CHCH .
   b. A saturated hydrocarbon is a compound consisting only of carbon and hydrogen with only single bonds between carbon atoms like methane, CH_4; ethane, CH_3CH_3; and pentane, $CH_3CH_2CH_2CH_2CH_3$.
   c. An unsaturated compound is one consisting only of carbon and hydrogen with one or more multiple bonds between carbon atoms like $H_3C-CH=CH-CH_2-CH=CH-CH_3$ where there are two carbon carbon double bonds.
   d. Alkenes are hydrocarbons with one or more carbon-carbon double bonds like $H_3C-CH_2-CH_2-CH_2-CH=CH-CH_2-CH_3$

9. In general metals will lose their valence electrons to form cations with a positive charge equal to the group number. Nonmetals will generally gain electrons to complete an octet.
   a. Bromine is in Group VIIA and has seven valence electrons, :Br· so it gains one to form the bromide ion, :Br:$^{1-}$
   b. Aluminum is in Group IIIA and has three valence electrons so it loses 3 to form Al^{3+}.
   c. Sodium is in Group IA and has one valence electrons so it loses one to form Na^{1+}.

d. Barium is in Group IIA and has two valence electrons so it loses two to form Ba^{2+}.
e. Calcium is in Group IIA and has two valence electrons so it loses two to form Ca^{2+}.
f. Ga is in Group IIIA like aluminum so it has three valence electrons; it loses these three electrons to form Ga^{3+}.
g. Iodine is in Group VIIA like Br so it has seven valence electrons; it forms the iodide ion, I^{1-}, by gaining one electron to complete the octet..
h. Sulfur is in Group VIA so it has six valence electrons, :S:̇ , and gains two electrons to form the sulfide ion, [:S:]²⁻ .
i. Group IA atoms lose one valence electron to form a +1 ion, see 9c above.
j. Group VIIA atoms have seven valence electrons and gain one to form a 1- ion, see 9g.

11. a. aluminum forms a 3+ ion and iodide forms a 1- ion
The charges on the ions can be used as subscripts on ions in the formula. The subscripts must be reduced to the smallest whole numbers.

This gives the formula as AlI_3

AlI_3 is named aluminum iodide.
b. strontium forms a +2 ion and chloride forms a -1 ion.

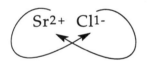   or Sr^{2+}
to balance the two + charges of 1 strontium ion

Cl^{1-} Cl^{1-}
two - charges are needed, so 2 chloride ions are in the formula

$SrCl_2$ is named strontium chloride.
c. Ca_3N_2 calcium nitride
d. K_2S potassium sulfide
e. Al_2S_3 aluminum sulfide
f. Li_3N lithium nitride

13. Answer will depend on individual search. Suggested categories: soft drinks, antacids, pickled items, breakfast cereals, preserved and smoked meats, breads, canned soups, jams and jellies, cleaning products.

15. a. Lithium is in Group Ia so it forms the Li^{1+} ion. Tellurium is in Group VIa and forms the Te^{2-} ion. The formula for the neutral ionic combination is Li_2Te. This means there are three ions in a formula unit, Li^+, Li^+ and Te^{2-} with a total charge of zero.
b. $MgBr_2$
c. Ga_2S_3

17. Generally the bond between a metal and a nonmetal is ionic and the bond between a nonmetal and a nonmetal is covalent.
    a. The bond between Na and S is ionic.
    b. The bond between N and Br is covalent.
    c. The bond between Ca and O is ionic.
    d. The bond between P and Br is covalent.
    e. The bond between C and O is covalent.

19. a. NO, nitrogen monoxide
    b. SO_3, sulfur trioxide
    c. N_2O, dinitrogen oxide
    d. NO_2, nitrogen dioxide

21. Ionic bonding exists between oppositely-charged ions. The ions are separate particles and no sharing of electrons occurs. The total charge on positive ions and negative ions totals up to zero. The polar covalent bond exists between atoms that share electrons unequally like the hydrogen and fluorine in HF. The nonpolar bonding exists between atoms that share electrons equally (atoms with similar attractions for electrons) like the hydrogens in H_2 or the carbons and hydrogens in hydrocarbons like CH_4.

23. a. The electron attracting power differences determine the polarity of the bonds. The farther apart two elements are in the periodic table the more polar the bond. The more nonmetallic an element, the greater its attraction for electrons. Elements to the right of the periodic table are the most nonmetallic. They attract electrons better. Carbon is to the right of hydrogen so carbon will attract electrons slightly better than hydrogen. Chlorine is even more to the right than carbon so it will attract electrons from carbon. The C-Cl bond is more polar because they are farther apart in the periodic table.
    b. Elements at the top of a group attract electrons better than elements lower in the same group. This means that fluorine will attract electrons better than chlorine. The C-F bond is more polar than the C-Cl bond.

25. Here the arrows are used to indicate the end of the covalent bond where the electrons tend to spend more time. When the atoms share electrons unequally, one end of the bond is slightly negative and the other end is slightly positive. The arrows indicate the direction of this shift.

    a. :C::::O:
       →
       Polar; O is the negative end and C is the positive end. The arrow indicates the direction electron shift in molecule.

    b. H:Ge:H (with H above and below) and tetrahedral Ge with 4 H's
       Nonpolar; the bonds are nonpolar. The molecule is tetrahedral and symmetric.

    c. Cl-B-Cl with Cl below, arrows pointing toward Cl's
       Nonpolar; the bonds are polar but the planar triangular shape puts the center of positive change and the center of negative charge in the middle of the boron atom.

    d. H:F:
       →
       Polar; F is the negative end and H is the positive end

27. a. The gaseous state has no definite shape or volume; a gas takes the shape and volume of its container. Gas molecules have high kinetic energy compared to the attractive forces between them and are therefore, free to move away from each other until the gas has filled the container. The distances between gas molecules are great, almost 1000 times the size of the molecules; this makes gases compressible.
    b. A liquid has definite volume and assumes the shape of the container. Molecules in a liquid are in contact with each other and attractions exist between them, keeping the liquid volume the same. The liquid molecules are free to move past each other, allowing the liquid to flow and assume the shape of its container. Because molecules are in contact already, there is very little room for them to move any closer together, and liquids are relatively incompressible.
    c. The solid state has definite shape and definite volume. The molecules or ions making up the solid are in direct contact and are occupying fixed positions in the solid arrangement, so shape and volume stay the same for a solid. The solid is incompressible due to direct contact between the particles.

29. solid, see answer to 27c

31. The number of gas molecules in a given volume of gas decreases as you move away from the earth's surface. Additionally, the force of earth's gravity decreases with altitude. This results in less weight or force per unit area at greater altitudes.

33. Gas molecules are constantly moving. The gas molecules of the perfume vapor diffuse away from the person wearing the perfume. This movement is random and the molecule can travel great distances.

35. Water has an unusually high normal boiling point, and surface tension. Additionally it expands on freezing. This latter property is different from most other compounds.

37. Evaporation of water from one's skin draws energy from the skin. This produces a resulting "chilling" sensation. The heat needed to make the water vaporize is provided by our body. We are cooled and the molecules gain energy. This same concept is used in evaporative air coolers and canteens that allow some water to pass out through the canteen walls to evaporate.

39. Without hydrogen bonding the boiling point for ammonia would be determined mainly by the molar mass of NH_3. The Group VA hydrides have the following boiling points SbH_3, -33°C; AsH_3, -62°C; and PH_3, -89°C. If the boiling point is plotted versus row in the periodic table, the projected boiling point for NH_3 without hydrogen bonding would be approximately -130°C. The actual boiling point for ammonia is much higher at -33°C.

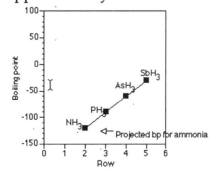

41. A gas is more soluble in a solvent at higher pressure because the increased pressure creates more collisions of gas molecules with the solvent surface. These collisions lead to more mixing of the gas molecules with the solvent and gas dissolving in the solvent. Champagne, sparkling waters, Coca Cola™ and Pepsi™ are examples of this process. When the gas pressure is decreased the solubility of the gas will decrease. This is what happens when the CO_2 pressure is released in an open bottle of soda.

43. The air in the frost-free refrigerator reaches an equilibrium between water molecules in the gas state and water molecules as frost. This equilibrium gas mixture is saturated with water molecules so it is "wet" or "moist". Dry air from the outside is blown into the refrigerator and the wet air is forced out. Then the gaseous water molecules and the frost (ice) reestablish equilibrium. The amount of frost decreases because some water molecules escape from the frost into the gas state. This process is repeated over and over again leading to the removal of frost in the refrigerator.

Speaking of Chemistry

Name _____

Chemical Bonding & States

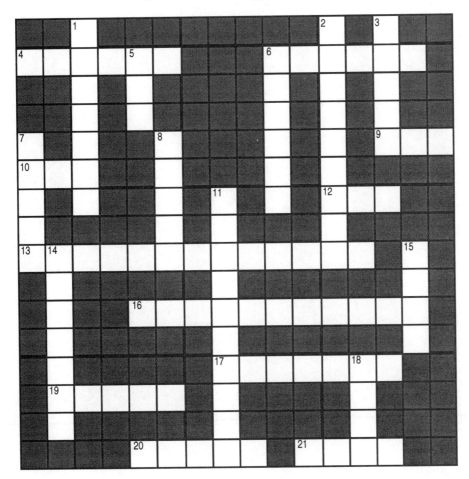

Across
4 A solvent dissolves a _____
6 Covalent bonding depends on ____ electrons
9 Number of electrons in a triple bond
10 Number of bonds formed by hydrogen
12 Prefix meaning three
13 Name for CO_2
16 Process of converting a solid to a vapor
17 Process in which liquid is converted to gas
19 Number of electrons in an octet
20 Name for O^{2-} ion
21 Water has a _____ shape

Down
1 Hydrocarbon containing a double bond
2 Hydrocarbon that has only single bonds between carbons
3 American scientist who explained bonding in terms of octet rule
5 Number of bonds normally formed by oxygen
6 Covalent bond where two electrons are shared
7 Type of bond formed in NaCl
8 Element in group IIIA with Z = 5
11 Covalent bond in CO
14 Hydrocarbon like acetylene
15 Prefix meaning one
18 Rare gas with symbol Ne
    fix meaning three.

© 1999 Harcourt Brace & Company. All rights reserved.

Bridging the Gap   I
"Hot Molecules": Kinetic Molecular Theory

The kinetic molecular theory is sometimes difficult to visualize. This activity is aimed at showing how temperature influences the motion and therefore the kinetic energy of molecules. According to the kinetic theory of matter, the molecules of a substance are in constant motion. The molecules move faster at high temperatures and slower at low temperatures. This can be clearly seen by looking at the rates of diffusion of food coloring dye molecules in common tap water. The average kinetic energy of a molecule is given by the equation

$$\overline{KE} = \frac{1}{2}m\overline{v}^2 = \frac{3}{2}RT$$

Here the average kinetic energy is $\overline{KE}$, the molar mass is m, the average velocity for the molecules is $\overline{v}^2$, the universal gas constant is R, the absolute temperature is T. Remember the absolute temperature equals the Celsius temperature plus 273.15. This equation predicts that the average kinetic energy will increase with temperature. The physical result of higher velocities is a shorter time between molecular collisions. These collisions are important in diffusion and dispersion. Each collision will knock a molecule around a little bit. If a clump of dye molecules is buffeted by surrounding solvent molecules, each collision will act to break up the clump. Eventually the clump will be dispersed and the mixture will be uniform with a homogeneous color.

Equipment and materials
Microwave oven or source of hot water, ice cubes or source of ice cold water, three identical clear colorless 12 ounce glasses, plastic tumblers or jelly jars, three sheets of white paper, tap water and a package of Schilling® or other brand of food coloring.

1. Fill one of the glasses with approximately 8 ounces of room temperature water.

2. Fill the second glass with an equal amount of ice cold water.

3. Heat some water in either a microwave oven or on a stove top burner. Fill the third glass with an equal amount of hot water.

4. Place each of the glasses in front of a sheet of white paper so the paper acts as a white background.

5. Stir each glass of water and then allow the three glasses to stand for about 10 minutes so the water is as still as possible and there is no mechanical movement.

6. Select one of the food coloring dyes.

7. Carefully add one drop of dye to each of the glasses, starting with the coldest and ending with the hottest. Observe how the drop of dye spreads throughout the water.

8. Note the approximate time required for the food coloring to be uniformly dispersed in any one of the glasses.

© 1999 Harcourt Brace & Company. All rights reserved.

Bridging the Gap   **I**   **Name** _____
"Hot Molecules": Kinetic Molecular Theory

Observations
What food coloring did you use?

Did the food coloring disperse at the same rate in the three different-temperature water samples? Which of the water samples dispersed the food coloring the fastest?

How did the drop of dye behave in the glass of room temperature water?

How did the drop of dye behave in the glass of hot water?

How did the drop of dye behave in the glass of ice cold water?

Analysis
What do you think causes the dye molecules in the food coloring to disperse the way they did in the three samples?

Do you believe you would observe the same results if you changed the food coloring? Justify your answer.

Do you believe you would see the same results if you used a different solvent like rubbing alcohol? Justify your answer.

Relation to Kinetic Molecular Theory
How do your observations relate to the Kinetic Molecular Theory?

Bridging the Gap II Name _____
Internet access to IBM Almaden Research Center:
Viewing Single Atoms with the Scanning Tunneling Microscope

The IBM Visualization Lab and Scanning Tunneling Microscopy
The physicists at IBM developed the scanning tunneling microscope to the point where it could be used to pick up and move individual atoms and ions. This work is so important that in 1986 the Nobel Committee awarded the Nobel Prize in physics to Gerd Binnig and Heinrich Rohrer for the development of the STM. This amazing instrument enables us to do what was once impossible, locate and work with single atoms and molecules.

Quantum Corrals
In this activity you are to open the IBM page at this URL and examine the "Quantum Corrals" image and discussion. You need to record the radius for the corral and the number of iron atoms used to make up the circumference of the circular corral.

http://www-i.almaden.ibm.com/vis/stm

What is the tabulated radius of the "Quantum Corral"? (1 Ångstrom = 1×10^{-10} meter)

How many iron atoms used to build the circle that makes up the corral?

The distance between iron atom centers can be estimated from the STM data given with image of the "Quantum Corral". You can calculate the circumference of the "corral" by using the following equation

$c = \pi(2r)$ where $\pi = 3.1416$; r = radius of circle

What is your value for the circumference of the "corral" in Ångstroms, (Å = 1×10^{-10} meter)
c =

The iron atoms are nested side by side to make up the "corral". If you divide the value for the circumference by the number of iron atoms you get an approximate value for the distance between iron atom centers.

$$d = \frac{\text{circumference of corral}}{\text{number of iron atoms}} = \frac{\text{Å}}{\text{iron atom}} = \text{_____}$$

© 1999 Harcourt Brace & Company. All rights reserved.

The tabulated diameter for iron atoms bonded in metallic iron crystals is 2.52 Ångstroms. Your calculated value is probably different.

Do you think the iron atoms in the corral are separated by some empty space? How much empty distance seems to exist between the iron atoms? Is this distance large compared to the size of an iron atom?
(Hint: Subtract the 2.52 Å from the distance you calculated per iron atom.)

If the iron atoms in the corral are merely set side by side do you think they are really close together enough to be "bonded" like the iron atoms in a piece of iron? Briefly justify your answer.

Bridging the Gap III
Odors and Molecular Shapes

There is evidence that the sense of smell is related to the geometry of molecules. The nose can easily tell the difference between a skunk and a rose. One theory suggests that seven primary odors are detected by specific receptors at the olfactory nerve ending; each primary odor is associated with a different receptor site. Five receptors are sensitive to shape. Two receptors are sensitive to electrostatic forces; one is attracted to molecules that have an electron rich region and the other is attracted to molecules that are have a localized positive charge, that is, a region that has a shortage of electron charge.

The seven primary odors are described as floral, musky, camphoraceous, pepperminty, ethereal, pungent and putrid. The shapes for the receptor sites are illustrated below. Example molecules are shown that would fit into the respective receptors. This theory explains why unrelated compounds can have the same or very similar odors: the geometric shapes or charged regions for the molecules are similar and they fit into the same receptors.

Complex odors result when more than one type of receptor site is stimulated.

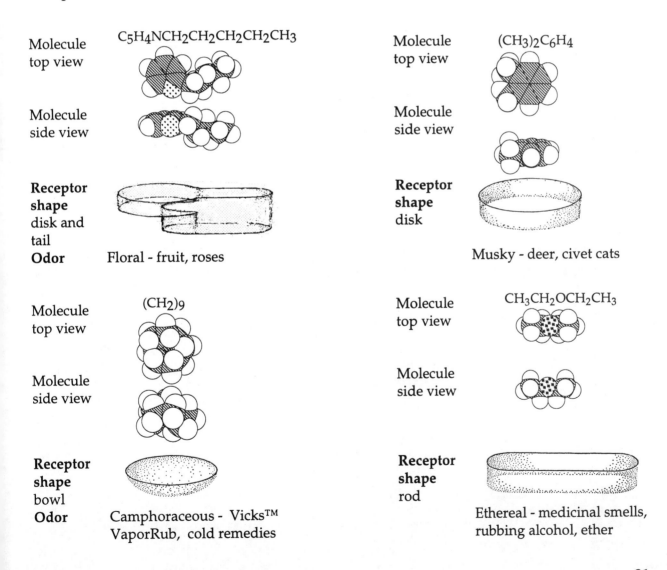

Molecule top view: $C_5H_4NCH_2CH_2CH_2CH_2CH_3$
Molecule side view
Receptor shape: disk and tail
Odor Floral - fruit, roses

Molecule top view: $(CH_3)_2C_6H_4$
Molecule side view
Receptor shape: disk
Musky - deer, civet cats

Molecule top view: $(CH_2)_9$
Molecule side view
Receptor shape: bowl
Odor Camphoraceous - Vicks™ VaporRub, cold remedies

Molecule top view: $CH_3CH_2OCH_2CH_3$
Molecule side view
Receptor shape: rod
Ethereal - medicinal smells, rubbing alcohol, ether

© 1999 Harcourt Brace & Company. All rights reserved.

Molecule top view	$CH_3C_6H_4(OH)CH(CH_3)_2$	Molecule top view	CH_3COOH	Molecule top view	CH_3CH_2SH	
Molecule side view	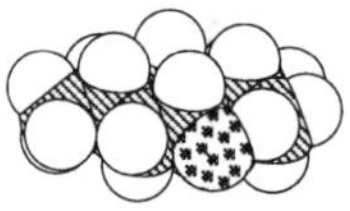	Molecule side view	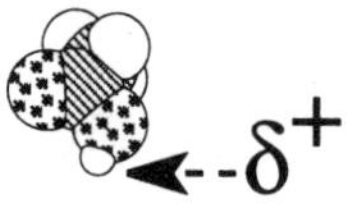 ←--δ^+	Molecule side view	←δ^-	
Receptor shape wedge	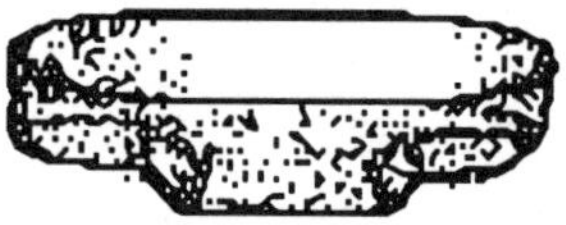	**Receptor has −charge**	 δ^-	**Receptor has + charge**	δ^+	
Odor	Pepperminty - toothpaste, Doublemint™ gum		Pungent - strong vinegar		Putrid - spoiled fish and meat	

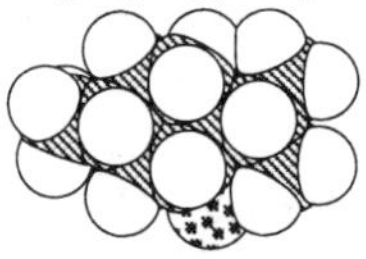

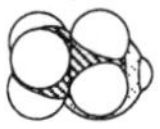

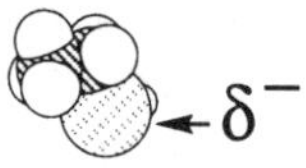

Bridging the Gap III Name _____

Odors and Molecular Shapes

The purpose of this exercise is to give you some practice in working with molecular shapes. You are supposed to predict the odor for each compound. You are to point out the structural features that make the molecule fit into the category. Identify the type of receptor (bowl, wedge, rod, etc.) for each molecule.

Matching Molecular Shapes, Receptor Types, and Odor

	$CH_3OCH_2CH_3$	$C_6H_7O(CH_3)_3$	$C_6H_5CH_2OCOCH_3$
Molecule top view			
Molecule side view			
Receptor site type			
Odor			

6   Chemical Reactivity: Chemicals in Action

6.1   Balanced Chemical Equations and What They Tell Us

This section describes how to read a chemical equation, and explains how the law of conservation of matter applies to chemical equations. The principles for balancing an equation are explained and illustrated with example reactions.

Objectives
After studying this section a student should be able to:
- describe the characteristics of a balanced equation
- describe the physical state of a reactant or product from parenthetical labels
- identify reactants and products in an equation
- distinguish a balanced equation from an unbalanced one
- balance an equation by inspection given the formulas for reactants and products
- explain why subscripts and formulas are not changed when balancing an equation
- describe the role of coefficients in a balanced equation
- identify the coefficients in a balanced equation
- state the law of conservation of matter

Key terms
reactants	products	balanced equation	coefficients
physical state	(s)	(g)	(ℓ)
(aq)	subscripts	conservation of matter	

6.2   The Mighty Mole and the "How Much?" Question

This section explains and illustrates how to count objects by weighing. The mole is defined relative to Avogadro's number, 6.022×10^{23}, and related to the molar mass. Examples are given showing how to compute the molar mass of a compound, 1 mol CO_2 = 44 g CO_2. The relationship between moles and chemical equations is explained and illustrated with examples. This section shows how to use a balanced chemical equation to determine mole ratios for products and reactants and predict the number of moles needed or produced. The molar mass is combined with the mole ratio to determine grams of reactants needed in a reaction and the grams of products formed.

Objectives
After studying this section a student should be able to:
- balance a chemical equation
- give a definition for molar mass
- describe the relationship between molar mass and Avogadro's number, 6.022×10^{23}
- calculate the molar mass for a compound using the formula and the periodic table
- examine a balanced equation and determine the mole ratio for any pair of substances
- calculate the expected grams of product formed when given the number of moles or grams of a reactant and the balanced equation
- calculate the number of grams of one reactant required to combine with a given number of grams of another and the balanced equation
- tell the moles of reactant needed and the numbers of moles of products formed in a reaction
- determine the number of grams of reactants needed in a reaction and the predicted number of grams of product

© 1999 Harcourt Brace & Company. All rights reserved.

Key terms

    Avogadro's number   molar mass   mole
    moles of reactant   moles of product   mass of reactant
    mass of product   mol ratio

6.3   Rates and Reaction Pathways: The "How Fast?" Question

This section discusses the path a reaction takes, rates of reaction, factors influencing reaction rates, activation energy and its relationship to reaction rate. The reasons why reaction rates can be controlled by adjusting temperature, concentration of reactants and addition of a catalyst are explained.

Objectives

After studying this section a student should be able to:
    state the relation between a reaction pathway and the rate of the process
    describe a reaction in terms of collisions between molecules
    state the role of a successful collision in the conversion of reactants to products
    give appropriate time periods for slow and fast reaction rates
    state definitions for reaction rate, activation energy, reaction pathway
    describe how molecule kinetic energy can influence the success or failure of a collision
    tell what the effect of a catalyst is on the rate of a reaction and on the activation energy
    describe how the rate of a reaction is influenced by temperature and concentration
    explain what role enzymes play in reactions

Key terms

    pathway   rate of a process   one-step
    kinetic energy   kinetic molecular theory   successful collision
    reaction rate   energy hill   fast reactions
    slow reactions   collisions per second   activation energy
    reaction direction   catalyst   effect of temperature
    effect of concentration   enzymes   uncatalyzed reaction

6.4   Chemical Equilibrium and the "How Far?" Question

This section explains what "dynamic equilibrium" means and describes chemical equilibrium for reversible reactions. Example reactions are the decomposition of limestone, $CaCO_3$, to produce lime, CaO; the ionization of acetic acid, CH_3COOH, and the synthesis of ammonia, NH_3. Le Chatelier's principle is used to explain how changes occur in equilibrium conditions and their effect on the reaction mixture.

Objectives

After studying this section a student should be able to:
    give a definition for dynamic equilibrium
    describe an example of a dynamic equilibrium
    explain what a reversible reaction means
    state Le Chatelier's principle and give an example of how it works
    explain what happens when a reaction proceeds to completion

Key terms

    dynamic equilibrium   chemical equilibrium   reversible
    Le Chatelier's principle   forward reaction   reverse reaction
    stress on a system   completion

6.5 The Driving Forces and the "Why?" Question

This section describes potential energy and ties it to endothermic and exothermic reactions. The concept of favorable and unfavorable reactions is explained. The first law of thermodynamics is described using examples. Definitions for the first law are summarized: Energy can be converted from one form to another but cannot be destroyed. The second law is explained in terms of entropy: The entropy of the universe is constantly increasing. Entropy is defined to be a measure of disorder. Entropy changes are linked to everyday events as well as chemical change.

Objectives

After studying this section a student should be able to:
- tell how a favorable reaction differs from an unfavorable reaction
- give the definition for endothermic reaction
- give the definition for exothermic reaction
- describe the relationship between entropy and disorder
- tell how reactions relate to entropy increase and entropy decrease
- explain how entropy content differs for solids, liquids, and gases
- state the first law of thermodynamics
- state the second law of thermodynamics

Key terms

favorable reaction	unfavorable reaction	potential energy
exothermic	endothermic	entropy
disorder	second law	first law of thermodynamics

6.6 Recycling: New Metals for Old

This section describes the problems associated with disposal of materials in landfills. The factors that affect the effectiveness and success of metals recycling are discussed. The recycling of metals like lead, copper aluminum zinc, iron and steel is discussed.

Objectives

After studying this section a student should be able to:
- describe how the amount of municipal solid waste has grown from 1960 to 1990
- tell which metal is the most recycled
- state the major sources for recycled steel and iron
- name the five most produced and valuable metals
- describe the merits of waste minimization instead of collecting, processing, and recycling

Key terms

solid waste	recycling plants	lead
recycled materials	market value	life span of products
cost of collection	cost of reprocessing	cost of disposal
environmental impact	quality of metal	factors in metal recycling

Additional Readings

Jones, Lynda. "Hot Gloves. (Spontaneous combustion by latex surgical gloves can cause fires)" Science World Nov. 3, 1995: 4.

Kingsbury, Donald. "The Janus-headed Arrow of Time: Entropy and Time Travel" Analog Science Fiction & Fact Feb. 1995: 58.

Zibel, Alan. "Recombinant Detergent" Popular Science Jan. 1996: 35.

© 1999 Harcourt Brace & Company. All rights reserved.

Answers to Odd Numbered Questions for Review and Thought

1. a. $CH_3CH_2OH_{(g)} + 3\ O_{2(g)} \longrightarrow 2\ CO_{2(g)} + 3\ H_2O_{(g)}$
   $1 \times (3+2+1) = 6$ H atoms $\qquad\qquad\qquad 3 \times (2) = 6$ H atoms
   In one molecule of CH_3CH_2OH or $CH_3CH_3O_1$ on the reactant side of the equation, there are 6 H atoms; this number is found by adding together the subscripts for H in the formula and multiplying by the coefficient, 1 (understood), in front of the ethanol formula in the balanced equation. The product side of the balanced equation also shows 6 H atoms; there are three water molecules on this side and each molecule contains 2 H atoms so the number of H's is found by multiplying the coefficient (3) by the subscript (2) for H in the water formula.

   b. $CH_3CH_2OH_{(g)} + 3\ O_{2(g)} \longrightarrow 2\ CO_{2(g)} + 3\ H_2O_{(g)}$
   Each side of this balanced equation shows 7 oxygen atoms, O. On the left there are 3 oxygen molecules, O_2, and each oxygen molecule has 2 oxygen atoms in it; there is one oxygen atom in the CH_3CH_2OH molecule; $3 \times 2 + 1 = 7$. On the right or product side there are 2 molecules of CO_2, and each contains 2 oxygen atoms; this accounts for 4 oxygen atoms. There are also 3 water molecules, each containing 1 oxygen atom; this accounts for 3 more O atoms. Therefore, on the right side of the equation there are $4 + 3 = 7$ oxygen atoms shown.

   c. $CH_3CH_2OH_{(g)} + 3\ O_{2(g)} \longrightarrow 2\ CO_{2(g)} + 3\ H_2O_{(g)}$
   The balancing coefficients are 1, 3, 2, and 3, respectively. The "1" in front of CH_3CH_2OH is understood to be there even though it usually is not written.

   d. The count for the number of atoms of an element is the same for both reactant and product sides of the equation. For example, in the equation above, there are 6 H atoms on the left of the arrow and 6 H atoms on the right side; there are 7 O atoms on each side and also 2 C atoms on each side.

3. a. To balance the equation $Al + Cl_2 \longrightarrow AlCl_3$ notice that the subscripts on the Cl are 2 and 3 on the left and right sides, respectively, of the arrow. Neither is a factor of the other, but this pair has a least common multiple of 6. The coefficient in front of Cl_2 must be "3" to give 6 Cl atoms on the left. The coefficient for $AlCl_3$ must be "2" to give 6 Cl atoms on the right. $Al + 3\ Cl_2 \longrightarrow 2\ AlCl_3$. Now turn to the aluminum, Al. On the right the count for Al is 2, because of the coefficient "2" in front of the $AlCl_3$. This means that 2 Al's must appear on the left side as well, so we need to put a "2" in front of the Al. The complete balanced equation is:
   $$2\ Al + 3\ Cl_2 \longrightarrow 2\ AlCl_3$$

   b. $Mg + N_2 \longrightarrow Mg_3N_2$ Notice that the subscript for N is a 2 in both reactant and product. This means that the coefficient for the N_2 will be the same as the coefficient for Mg_3N_2; the smallest possible coefficient is "1". $Mg + 1\ N_2 \longrightarrow 1\ Mg_3N_2$ Now count the magnesium atoms. The product has $1 \times 3 = 3$ Mg atoms. To equal this we put a "3" in front of the Mg on the left of the equation.
   $$3\ Mg + N_2 \longrightarrow Mg_3N_2$$

c. NO + O_2 ---> NO_2 Examine the equation and count the number of O atoms on each side and the number of N atoms on each. There are 2 oxygen atoms in NO_2 so whatever coefficient we place in front of this in the equation will give us an even number of oxygen atoms on the right; the number of oxygen atoms on the right will be 2 or 4 or 6, etc. The left side already shows 1 + 2 = 3 oxygen atoms so we know we can't get a balanced equation with only 2 oxygens on the right. This leads us to try a "2" in front of the NO_2 to give 4 oxygen atoms and 2 nitrogen atoms on the right. To get 2 nitrogen atoms on the left we need a "2" in front of the NO; this will also account for 2 oxygen atoms in the two molecules of NO. Adding the 2 oxygen atoms shown in the 2 NO's to the 2 oxygen atoms shown in one O_2 molecule gives 4 oxygen atoms on the left.

$$2 \text{ NO} + 1 \text{ } O_2 \longrightarrow 2 \text{ } NO_2$$

d. SO_2 + O_2 ---> SO_3 The SO_2 and SO_3 molecules each contain one S atom so we know the coefficients for SO_2 and SO_3 must be the same. We also know that a "1" will not work for this coefficient because that would give 3 oxygen atoms on the right and we can already see 4 oxygens on the left. So, if the coefficient "1" will not work, we next try a "2". Trying a "2" in front of SO_3 and a "2" in front of SO_2 balances the S atoms. Next, looking at the oxygen atoms we see 2 x 3 or 6 oxygen atoms on the right; we see 2 x 2 oxygens in the 2 SO_2's on the left and another 2 oxygens in the single O_2 molecule, so we can account for 6 oxygen atoms on the left.

$$2 \text{ } SO_2 + O_2 \longrightarrow 2 \text{ } SO_3$$

e. H_2 + N_2 ---> NH_3 This equation is balanced by seeing that the subscripts for hydrogen are a 2 in the H_2 and a 3 in the ammonia, NH_3. The lowest common multiple of 2 and 3 is 6 so the coefficient for the H_2 must be a "3" and the coefficient for the NH_3 must be a "2". Next, checking the nitrogen atoms we see 2 N's in the 2 NH_3 molecules on the right. For 2 N's on the left we need one N_2 molecule

$$3 \text{ } H_2 + N_2 \longrightarrow 2 \text{ } NH_3$$

5. a. $Ba_{(s)}$ + $H_2O_{(\ell)}$ ---> $Ba(OH)_{2(aq)}$ + $H_{2(g)}$ This reaction is balanced more easily when the break up of the H_2O molecule into OH and H is recognized. This often happens when water is a reactant. This means that each OH among the products came from one water molecule or that every OH in products must be matched by an H_2O molecule in the reactants. To get 2 OH units in the $Ba(OH)_2$ product requires that 2 H_2O molecules react. If 2 water molecules break apart, there will also be 2 H's produced that can join to make one H_2 molecule. The coefficients for Ba and $Ba(OH)_2$ must be the same because the subscripts for Ba are the same in both.

$$Ba_{(s)} + 2 \text{ } H_2O_{(\ell)} \longrightarrow Ba(OH)_{2(aq)} + H_{2(g)}$$

b. $Fe_{(s)} + H_2O_{(\ell)} \longrightarrow Fe_3O_{4(s)} + H_{2(g)}$ Note that the oxygen atoms in the Fe_3O_4 product must come from the H_2O molecules on the left. To get 4 oxygens on the right we pick a coefficient of "4" to place in front of the H_2O.
$Fe_{(s)} + 4\ H_2O_{(l)} \longrightarrow Fe_3O_{4(s)} + H_{2(g)}$ The subscript for the H is the same in both H_2O and H_2 and hydrogen appears only in these two molecules, so the coefficients for H_2O and H_2 must be the same (4). Alternatively, we could look at the 4 H_2O on the left and recognize that there are 8 H's there. To get 8 H's on the right requires a coefficient of "4" in front of the H_2. Lastly, there must be 3 Fe's on each side so we put a "3" in front of the Fe on the left.
$$3\ Fe_{(s)} + 4\ H_2O_{(l)} \longrightarrow Fe_3O_{4(s)} + 4\ H_{2(g)}$$

c. $Na_{(s)} + H_2O_{(l)} \longrightarrow NaOH_{(aq)} + H_{2(g)}$ This reaction is similar to 5a. The H_2O molecules break apart to yield OH and H. It is helpful to notice that the coefficients for Na and NaOH must be the same because one Na is shown is each; the coefficients for H_2O and NaOH must be the same because the subscripts on the oxygen are the same; the coefficients for H_2O and H_2 are the same because the subscripts on H are the same. If we recognize that one H comes from the break up of one water molecule, then we can conclude that 2 water molecules are needed to give up the 2 H's shown in the H_2 molecule. We put a "2" in front of the H_2O. This will also produce 2 OH's so we need a "2" in front of the NaOH to account for 2 OH's. This then shows 2 Na's on the right and means we must put a "2" in front of the Na on the left.
$$2\ Na_{(s)} + 2\ H_2O_{(\ell)} \longrightarrow 2\ NaOH_{(aq)} + H_{2(g)}$$

d. This equation is balanced like 5 c. except that Li is reacting in place of Na. This shows how all the elements of Group IA react with water.
$$2\ Li_{(s)} + 2\ H_2O_{(\ell)} \longrightarrow 2\ LiOH_{(aq)} + H_{2(g)}$$

7. a. $Sn_{(s)} + HBr_{(aq)} \longrightarrow SnBr_{2(aq)} + H_{2(g)}$ Note that the subscript for Br in HBr is a 1 and the subscript for Br in $SnBr_2$ is a 2. The lowest common multiple is 2 x 1 = 2; the coefficient for the HBr must be a "2" if the coefficient for $SnBr_2$ is a "1".
$$Sn_{(s)} + 2\ HBr_{(aq)} \longrightarrow SnBr_{2(aq)} + H_{2(g)}$$
This also shows one Sn and 2 H's on each side of the arrow so the equation is balanced.

b. $Mg_{(s)} + HCl_{(aq)} \longrightarrow MgCl_{2(aq)} + H_{2(g)}$ This equation is balanced the same way as 7 a. except that Mg is reacting instead of Sn and HCl is used in place of HBr. Note that all Group IIA metals will react in a similar way with HCl.
$$Mg_{(s)} + 2\ HCl_{(aq)} \longrightarrow MgCl_{2(aq)} + H_{2(g)}$$

c. This equation is balanced in the same way as 5 a. Ba and Ca are both from Group IIA and react similarly. $Ca_{(s)} + 2\ H_2O_{(\ell)} \longrightarrow 2\ Ca(OH)_{2(aq)} + H_{2(g)}$

d. $Zn_{(s)} + HNO_{3(aq)} \longrightarrow Zn(NO_3)_{2(aq)} + H_{2(g)}$ This equation is balanced more easily if we recognize that the HNO_3 molecule comes apart into NO_3 and H, much as the Cl and H separate in the HCl in 7 a. To find 2 NO_3 units in the product $Zn(NO_3)_2$ requires the break up of 2 HNO_3's, so we put a "2" in front of the HNO_3 on the left. Break up of 2 HNO_3's also frees 2 H's which can go together to make one H_2 molecule. Since there is one Zn on the right, we only need one Zn on the left.
$$Zn_{(s)} + 2\ HNO_{3(aq)} \longrightarrow Zn(NO_3)_{2(aq)} + H_{2(g)}$$

e. This equation is balanced just like 5 c and 5 d except that cesium, Cs, is the reactant metal instead of Na or Li. Note that Li, Na, and Cs are all members of Group IA in the periodic table. $2\ Cs_{(s)} + 2\ H_2O_{(\ell)} \longrightarrow 2\ CsOH_{(aq)} + H_{2(g)}$

9. The carbon-12 isotope with a <u>defined</u> mass of <u>exactly</u> 12 amu is the basis for the atomic mass scale. The element carbon has an average atomic weight greater than 12.0000 because some naturally-occurring isotopes of carbon are heavier than the carbon-12 isotope. This makes the atomic weight larger than defined 12 amu since the periodic table value atomic weight shows the weight of an *average* carbon atom.

11. Each is 0.500 mol of that element. 1 mol Pb = 207.0 g Pb and 1 mol C = 12.012 g C.

$$\frac{103.5\ g\ Pb}{1} \times \frac{1\ mol\ Pb}{207.0\ g\ Pb} = 0.5000\ mol\ Pb \qquad \frac{6.006\ g\ C}{1} \times \frac{1\ mol\ C}{12.012\ g\ C} = 0.5000\ mol\ C$$

13. Avogadro's number is 6.022×10^{23} to 4 significant digits or 6.02×10^{23} to 3 s.d.

15. The mol is defined as the number of atoms in exactly 12 g of carbon-12, about 6.022×10^{23} atoms.

17. Some reactions are fast because reactants have weak bonds. These reactions have low activation energies. (These reactions have a low energy barrier or a "low hill to climb" for reactants to form products.) Reactions between gas phase molecules occur rapidly because collisions between reactant molecules are more frequent for gases. Some reactions are slow because reactants have strong bonds and a high activation energy. These reactants have a high "energy hill to climb" to form products.

19. Freezing slows reaction rates; this slows spoilage and slows the growth of mold or bacteria. The lower temperature decreases the kinetic energy of reactant molecules, causing them to move more slowly. This decreases the chance of a collision between reactants. Also the collisions will have low energies when they do occur, so the chance of having the necessary activation energy is less. The freezing process tends to immobilize molecules in a solid where the freedom of movement is very low, diminishing chances for collisions even more. This means that molds and bacteria grow much more slowly.

© 1999 Harcourt Brace & Company. All rights reserved.

21. Reactions with high or large activation energies (examples I and II) are slower than reactions with low or small activation energies (examples III and IV).

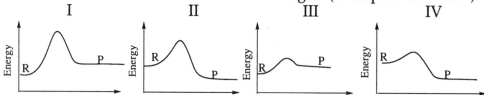

23. The hydrogen and oxygen will not react at room temperature because the molecules do not have the required activation energy. The activation energy is high (large) so there is no reaction at room temperature. The spark provides the needed activation energy to start the reaction. The reaction is exothermic and the energy released provides the energy needed to keep the reaction going; the reaction occurs quickly since it has a "built-in" source of energy. Any "extra" energy is released to the surroundings. A reaction produces energy because the products have lower potential energy than the reactants.

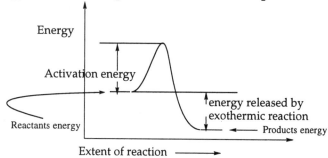

25. Burning paper requires that the molecules in the paper react with O_2 molecules. Paper (and other substances) burn more rapidly in pure oxygen because collisions between oxygen molecules and the paper are more numerous or more frequent if oxygen is the only gas present. Air is about 80% nitrogen gas and only about 20% oxygen gas, so in air only about 20% of the collisions between the paper and the gas molecules will involve an O_2 molecule and contribute to the reaction. When N_2 molecules collide with the paper, there will be no reaction.

27. Reversible reactions occur in both forward and reverse directions. Consider the reaction $N_2 + 3 H_2 \rightleftharpoons 2 NH_3$ The double arrow indicates reversibility. This means that both the reaction $N_2 + 3 H_2 \longrightarrow 2 NH_3$ and the reaction $2 NH_3 \longrightarrow N_2 + 3 H_2$ can and do occur at the same time. When the reaction rates for both of these reactions are exactly the same, equilibrium exists between the forward and reverse processes. As fast as 2 ammonia molecules form, somewhere else in the reaction mixture 2 ammonia molecules are being converted to 1 nitrogen and 3 hydrogen molecules. Once equilibrium is established between forward and reverse processes, the net amount of each substance in the reaction mixture stays the same.

29. When limestone is roasted the reaction is $CaCO_{3(s)} \rightleftharpoons CaO_{(s)} + CO_{2(g)}$
    a. If CO_2 is added to the reaction mixture, the reaction will shift toward $CaCO_{3(s)}$. This happens because the added CO_2 will initially collide more often with the CaO and increase the rate of formation of $CaCO_3$. This higher rate will continue until a new balance for equilibrium comes into being. The new equilibrium will involve slightly more CO_2, less CaO, and more $CaCO_3$ than the original equilibrium balance.

b. If carbon dioxide is allowed to escape, the reaction will shift to replace the missing CO_2. This means that the rate of disappearance of $CaCO_{3(s)}$ will initially increase as CO_2 and CaO form until a new equilibrium is established. The new equilibrium will contain less $CaCO_3$, more CaO and slightly less CO_2 that the original.

31. When a reversible reaction shifts to favor products, additional reactants are consumed to make more products until a new balance is reached.

33. After the reaction of HCl and NaOH has gone to completion, the reactant concentrations are practically zero; essentially, all reactants have been converted to products. The reaction is considered to be essentially nonreversible.

37. a. Potential energy is stored energy or energy due to the positions of particles. A book on a shelf has potential energy relative to the floor below. Two charged particles have potential energy due to the attraction (opposite charges) or repulsion (like charges) between them.
   b. An exothermic process gives off energy. This occurs because the bonds in the products store less energy than the bonds in the reactants. High energy relatively unstable bonds are replaced by more stable lower energy bonds. The energy released equals this energy difference between the two sets of bonds.
   c. An endothermic process consumes or takes in energy. This occurs because the bonds in the products store more energy than the bonds in the reactants. Low energy relatively stable bonds are replaced by more unstable high energy bonds. The energy required equals the energy needed to break the old bonds and form the new higher energy ones..
   d. Entropy is a measure of chaos or disorder. A well-ordered deck of cards has low entropy and a shuffled deck has higher disorder and entropy. Natural processes tend to go to higher entropy. Therefore, energy and work must be used to reduce the entropy of these systems. The observed fact is that all naturally occurring changes are accompanied by an increase in entropy for the system and surroundings. Local decreases are off-set by increases elsewhere in the universe. The entropy of the universe is increasing. This value for the universe is the combination of changes for any system plus the change for the rest of the universe, the surroundings.
   e. A favorable reaction is a process that favors products at the expense of reactants. The reaction is described as going to completion.

37. You can write a book answering this question! The symbol for entropy is "S". A change in entropy is represented by ΔS. Every spontaneous process occurs with an increase in entropy for the universe. The total entropy change for the universe is the sum of the entropy change for the system being examined plus the entropy change for the surroundings: $\Delta S_{universe} = \Delta S_{system} + \Delta S_{surroundings}$. This entropy change for the universe consumes energy that cannot be used for other purposes. Every spontaneous process has a built-in amount of this so called "wasted" energy.

39. Burning methane is an exothermic reaction. Energy is released by the molecules in the reaction. We observe this when we notice that burning methane or natural gas produces heat:
$CH_4 + 2 O_2 \longrightarrow CO_2 + 2 H_2O + heat$.

41. No, it is not favorable because it requires a continuous input of energy to keep the process going; without this energy the process will stop. The entropy decrease for building up a large molecule from several small ones is accompanied by a larger entropy increase in the surroundings. The total entropy of the universe increases:
$\Delta S_{universe} = \Delta S_{photosynthesis} + \Delta S_{surroundings}$ = + value. The chaos created in the surroundings exceeds the organization created in making the large glucose molecule.
$6\ CO_2 + 6\ H_2O \longrightarrow C_6H_{12}O_6 + O_2$

43. No, recycling of waste does not violate the 2nd law by creating order in the system of waste material. The energy used to organize the recycled material creates disorder in the surroundings. The resulting total entropy change is an increase in disorder. This is illustrated by the production of entropy when people use energy to sort and collect recyclables. $\Delta S_{universe} = \Delta S_{recycled\ waste} + \Delta S_{sorting,\ collecting,\ processing}$

Solutions for Problems

1. The molar mass is calculated using the formulas and the atomic weights from the periodic table. The first step is to count the number of atoms of each element in the formula; this tells the number of mols of each element needed for one mol of the compound. Masses are rounded to give whole numbers for ease of calculation; in other problems masses will be rounded to the number of significant digits appropriate to match the factors or terms in the rest of the problem.

    a. Atomic weight of H = 1.01. Atomic weight of O = 16.00
    One molecule of water contains 2 atoms of hydrogen and 1 atom of oxygen.
    One *mol* of water contains 2 *mols* of hydrogen and 1 *mol* of oxygen.

    $2\ mol\ H = 2\ g\ H \qquad 2\ mol\ H \times \dfrac{1.01\ g\ H}{1\ mol\ H} = 2.02\ g\ H = 2\ g\ H$

    $1\ mol\ O = 16\ g\ O \qquad 1\ mol\ O \times \dfrac{16.00\ g\ O}{1\ mol\ O} = 16.00\ g\ O = 16\ g\ O$

    $1\ mol\ H_2O = 2 + 16 = 18\ g\ H_2O$

    b. 1 molecule I_2 contains 2 atoms I; 1 mol I_2 contains 2 mols I. Atomic Weight I = 127
    molar mass I_2 = mass of 1 mol of I_2 = $2\ mol\ I \times \dfrac{127\ g\ I}{1\ mol\ I} = 254\ g\ I_2$

    c. 1 mol KOH contains 1 mol K, 1 mol O and 1 mol H.
    molar mass KOH = 39 g K + 16 g O + 1 g H = 56 g KOH

    d. 1 mol NH_3 contains 1 mol N and 3 mols H
    molar mass NH_3 = 14 g N + (3 mols H)(1 g H/ 1 mol H) = 17 g NH_3

    e. 1 mol CO_2 contains 1 mol C and 2 mols O
    molar mass CO_2 = 12 g C + (2 mol O)(16 g O/ 1 mol O) = 12 + 32 = 44 g CO_2

    f. 1 mol CO contains 1 mol C and 1 mol O
    molar mass CO = 12 g C + 16 gO = 28 g CO

2. a. 1 mole $C_6H_{12}O_6$ = 6 × 12.0 g C + 12 × 1.0 g H + 6 × 16.0 g O
   = 72.0 g C + 12.0 g H + 96.0 g O = 180.0 g $C_6H_{12}O_6$
   b. 1 mole H_2SO_4 = 2 × 1.0 g H + 1 × 32.0 g S + 4 × 16.0 g O
   = 98.0 g H_2SO_4

   c. 1 mole Na_2HPO_4 = 2 × 23.0 g Na + 1 × 1.0 g H + 1 × 31.0 g P + 4 × 16.0 g O
   = 142.0 g Na_2HPO_4

   d. 1 mole $Ca(NO_3)_2$ = 1 × 40.1 g Ca + 2 × 14.0 g N + 6 × 16.0 g O
   = 40.1 g Ca + 28.0 g N + 96.0 g O = 164 g $Ca(NO_3)_2$

   e. 1 mole $C_{57}H_{104}O_6$ = 57 × 12.0 g C + 104 × 1.0 g H + 6 × 16.0 g O
   = 684.0 g Ca + 104.0 g H + 96.0 g O = 884.0 g $C_{57}H_{104}O_6$

3. Look up the Atomic Weight of each element in the periodic table. This also tells the mass in grams of 1 mol of the element. "Molar mass" means the mass of 1 mol of substance.
   a. 1 mol Cu = 63.55 g Cu;   b. 1 mol Pb = 207.2 g Pb;   c. 1 mol Na = 22.99 g Na.
   The molar mass of lead, Pb, is the largest.

4. You must determine the molar mass for each substance and multiply by the number of moles. mass = (# moles) × (# grams / mole)
   a. # grams CO_2 = $\left[\dfrac{0.5 \text{ mole } CO_2}{1}\right]\left[\dfrac{44.0 \text{ g } CO_2}{1 \text{ mole } CO_2}\right]$ = 22.0 g CO_2

   b. # grams Li = $\left[\dfrac{2.0 \text{ mole Li}}{1}\right]\left[\dfrac{7.0 \text{ g Li}}{1 \text{ mole Li}}\right]$ = 14.0 g Li

   c. # grams H_2 = $\left[\dfrac{12.0 \text{ mole } H_2}{1}\right]\left[\dfrac{2.0 \text{ g } H_2}{1 \text{ mole } H_2}\right]$ = 24.0 g H_2

   12 moles of hydrogen has the largest mass

5. a. Cu(s) + Cl_2(g) ---> $CuCl_2$(s)

   1 mol Cu   1 mol Cl_2 forms   1 mol $CuCl_2$

   The balanced equation gives the following mol relationships.

   1 mol Cu per 1 mol Cl_2 or 1 mol Cl_2 per 1 mol Cu

   # mol copper = $\left[\dfrac{0.5 \text{ mol } Cl_2}{1}\right]\left[\dfrac{1 \text{ mol copper}}{1 \text{ mol } Cl_2}\right]$ = 0.5 mol copper

   b. Assuming an unlimited amount of Cl_2 the moles of $CuCl_2$ can be calculated.

   1 mol $CuCl_2$ per 1 mol Cu; # mol = $\left[\dfrac{1.5 \text{ mol Cu}}{1}\right]\left[\dfrac{1 \text{ mol } CuCl_2}{1 \text{ mol Cu}}\right]$ = 1.5 mol $CuCl_2$

6. The first step is to balance the equation:
$$C_6H_{12}O_{6(aq)} \longrightarrow 2\ CH_3CH_2OH_{(aq)} + 2\ CO_{2(g)}$$

The molar interpretation can be made from the coefficients in the balanced equation.

$C_6H_{12}O_{6(aq)}$	$\longrightarrow$	$2\ CH_3CH_2OH_{(aq)}$	+	$2\ CO_{2(g)}$
1 mole $C_6H_{12}O_6$		2 moles CH_3CH_2OH		2 moles CO_2

a. There are two ways to approach the solution to this problem. You can consider what happens to the amounts of products if the amount of glucose is 6 times as much as the amount of glucose shown in the balanced equation; if 6 moles of glucose react, the amounts of ethanol and carbon dioxide will also be multiplied by 6.

glucose		ethanol		carbon dioxide
$C_6H_{12}O_{6(aq)}$	$\longrightarrow$	$2\ CH_3CH_2OH_{(aq)}$	+	$2\ CO_{2(g)}$
1 mole $C_6H_{12}O_6$		2 moles CH_3CH_2OH		2 moles CO_2
6 x 1 mole $C_6H_{12}O_6$		6 x 2 moles CH_3CH_2OH		6 x 2 moles CO_2
6 mole $C_6H_{12}O_6$		12 moles CH_3CH_2OH		12 moles CO_2

The short method to solve this problem is to determine the mole-ratio between glucose and ethanol from the balanced equation and use this mole-ratio as a conversion factor.

$$\frac{1\ mol\ C_6H_{12}O_6}{2\ mol\ CH_3CH_2OH} \quad or \quad \frac{2\ mol\ CH_3CH_2OH}{1\ mol\ C_6H_{12}O_6}$$

$$6\ mols\ C_6H_{12}O_6 \times \frac{2\ mol\ CH_3CH_2OH}{1\ mol\ C_6H_{12}O_6} = 12\ mols\ CH_3CH_2OH$$

b. The moles of ethanol can be determined from moles of carbon dioxide using the mole-ratio that comes from the balanced equation: 2 moles ethanol/ 2 moles carbon dioxide.

$$10.5\ mols\ CO_2 \times \frac{2\ mol\ CH_3CH_2OH}{2\ mol\ CO_2} = 10.5\ mols\ CH_3CH_2OH$$

Speaking of Chemistry

Chemical Reactivity

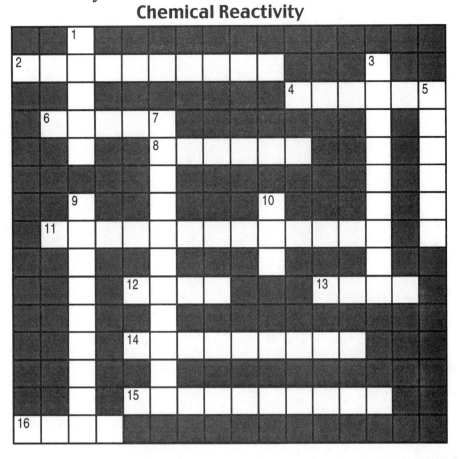

Across
2 Process that releases heat.
4 Change applied to an equilibrium system.
6 The ____ law of thermodynamics, predicts conservation of energy.
8 A biological catalyst.
11 The strength of solutions and mixtures are measured in _____ units.
12 The speed of reaction and change of concentration with time indicate the _____ .
13 Reaction rate is_____when reactants are consumed quickly.
14 The number 2 and 3 in the Fe_2O_3 formula are called_____.
15 Process that proceeds in both the forward and reverse directions.
16 An Avogadro's number of formulas.

Down
1 State in which particles are in contact and in fixed positions.
3 Materials on the left hand or beginning side of an equation
5 The _____ law of thermodynamics says entropy is always increasing.
7 A _____ increase will raise the kinetic energy of particles.
9 Stored energy.
10 Physical state where particles are far apart.

Bridging the Gap   I   Name _____
Sodium Bicarbonate, NaHCO₃, and Vinegar
Heat Effects and Production of CO₂

This exercise will deal with the reaction between vinegar and baking soda. Vinegar is a mixture that is roughly 95% water and 5 % acetic acid, CH_3COOH. Baking soda is pure sodium bicarbonate, $NaHCO_3$. The reaction is shown below.

Acetic acid	Baking soda		Water	Carbon dioxide	Sodium acetate
$CH_3COOH(aq)$	+ $NaHCO_3(s)$	$\longrightarrow$	$H_2O(l)$	+ $CO_2(g)$	+ $Na^+CH_3COO^-(aq)$

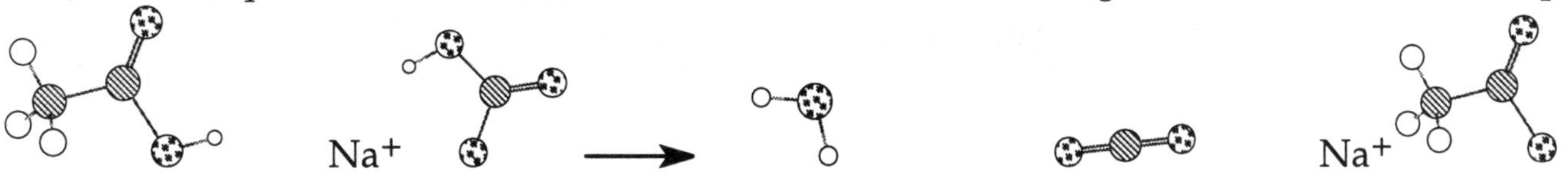

Heat Effects
One part of this activity is to observe the heat effect that accompanies the reaction. This will allow you to decide if the reaction is exothermic or endothermic. When a reaction is exothermic the reaction releases energy in the form of heat or light to the surroundings. A fire is a reaction that produces both heat and light. Sometimes this energy release is only in the form of heat. The surroundings gain energy (from the reaction) and become warmer. You can observe this by touching the container. When a reaction is endothermic the reaction draws energy from the surroundings. In this case the surroundings lose energy and create a "cooling" sensation if you touch the reaction container.

Production of Carbon Dioxide, CO₂(g)
This part of the exercise is to observe the reaction between vinegar and baking soda and note the amount of gas, CO_2, produced.

Equipment and materials
A box of Ziploc® sandwich bags (or equivalent), a roll of Saran® or other plastic wrap, a bottle of plain white vinegar, a box of baking soda (typically Arm and Hammer®), a 1/4 cup measuring cup, 1/8 and 1/4 teaspoon measuring spoons, scissors, marking pen, and transparent tape or stick on labels.

© 1999 Harcourt Brace & Company. All rights reserved.

Procedure

1. Label three Ziploc® bags as 1, 2, and 3. Fold back the top of the open Ziploc® bag back so the bag stay open. Do this with all three bags. Set the bags in a bowl or cup so they stand up and don't spill.

2. Add 1/4 cup of vinegar to each one of the three Ziploc® bags.

3. Open the box of baking soda and stir the contents so that samples are not taken only from the surface which may have reacted or decomposed.

4. Cut three pieces of plastic wrap about 3x3 inches each. These will be used to wrap the sodium bicarbonate (baking soda).

5 Layout the three pieces of plastic wrap on a clean surface. Measure out the baking soda with the measuring spoon. You can mark the plastic wrap with a marking pen to label the samples 1, 2, 3 if you wish.

Sample 1	Sample 2	Sample 3
1/8 teaspoon of baking soda	1/4 teaspoon of baking soda	1/2 teaspoon of baking soda

6. Fold up the plastic wrap around the baking soda samples. This will allow you prevent the mixing of the vinegar with the baking soda until you are ready for the reaction to occur.

7. Carefully place "Sample 1" of the wrapped sodium bicarbonate into the bag labeled "1". Try to "float" the wrapped sample on the surface of the vinegar. Carefully seal the bag. Be sure the bag is sealed completely, but do not allow the two reactants to mix yet..

8. From the outside of the sealed Ziploc® bag put your fingers on the package of baking soda. Rub the package with your fingers so the it opens up. Shake the Ziploc® bag so the contents mix. Immediately, some bubbles should form. Watch what happens. Touch the outside of the bag to sense its temperature. Record your observations on the Concept Report Sheet. KEEP this bag sealed and set it aside.

9. Repeat steps 7 and 8 with the other Ziploc® bags.

10. Estimate the amount of gas produced in each bag by carefully rolling the top of the bag down so the gas is trapped in the bottom of the bag. Stop rolling when you meet resistance. The skin of the bag will tighten and look like a balloon. Estimate the relative volumes of the three bags. (1/4 of a bag full, 1/2 bag full, etc.)

11. Flush the contents of the bags down the drain with a steady flow of water.

Bridging the Gap I Name_____
Sodium Bicarbonate, NaHCO$_3$, and Vinegar, CH$_3$COOH
Heat Effects and Production of CO$_2$

Observations
Record the heat effects and volume changes you observed when you mixed the sodium bicarbonate with the vinegar.

Mixture	Gas volume observation (A fourth of a bag, etc.)	Temperature observation (No effect, warmer, cooler)
Vinegar and 1/8 teaspoon of baking soda		
Vinegar and 1/4 teaspoon of baking soda		
Vinegar and 1/2 teaspoon of baking soda		

Analysis and Conclusions

1. Is the reaction exothermic or is it endothermic? Justify your answer based on your observations.

2. What happened to the volume of the gas in the Ziploc® bag when you increased the amount of sodium bicarbonate used? Why do you think this happened?

3. Write the overall reaction between vinegar and sodium bicarbonate. Circle the reaction product you believe causes the observed volume changes?

4. What do you think would happen if you continued to add larger and larger amounts of baking soda? Do you think the volume would continue to increase or would there be some limit? Why?

Bridging the Gap   II   Name _____
The Formation of Ammonia, NH₃
Fritz Haber, The Nobel Prize in Chemistry 1918, and The Haber Process

The number of chemical reactions that are possible are almost endless; however, a few of these reactions are extremely important to the human race. The chemical synthesis of ammonia from hydrogen and nitrogen is one of these vital reactions.

Nobel Prize in Chemistry 1918
Fritz Haber was awarded the 1918 Nobel Prize in Chemistry for his method of synthesizing ammonia from its elements, nitrogen and hydrogen. The need for nitrogen fertilizers is described at this site.
    http://www.qpais.co.uk/nable/whycare.htm

1. Why is the synthesis of NH_3 from H_2 and N_2 so wonderful that it merited a Nobel Prize?

The Nobel Foundation in Sweden maintains this site. You can access many details of Haber's work here. You can get answers to the following questions at this URL.
    http://www.nobel.se/laureates/chemistry-1918-press.html

2. In 1904 Haber and van Oordt showed that ammonia, NH_3, could be formed at about 1000°C and high pressures. What is the balanced equation for the synthesis of ammonia?

3. What is the approximate temperature for the reaction when it is done at "dark red heat"?

4. Haber needed to use catalysts to make his synthesis reaction happen. Catalysts are materials that facilitate a reaction, but are not consumed in the process. Haber used iron as a catalyst for some of his work, but found that other metals were better. Name two other catalysts that worked better than iron. Would you class these materials as metals or nonmetals?

5. When land is intensively cultivated, what happens to the amount of nitrogen-containing compounds in the soil?

Optional
This site by Unocal Agricultural chemicals and gives answers about agricultural ammonia use.
   http://www.uncal.com/agriproducts/products/anhydrs.htm

6. How is ammonia used in food production?

7 Acid - Base Reactions

7.1 Acids and Bases: Chemical Opposites

Main topics in this section are the properties of acids, the definition of acids as proton donors, the definition of bases as hydroxide donors and as proton acceptors, the behavior of acids in aqueous solutions, equations illustrating reactions of acids with water, the behavior of bases in aqueous solutions, and equations illustrating neutralization reactions.

Objectives
After studying this section a student should be able to:
- describe the characteristic properties of acids
- describe the characteristic properties of bases
- describe the effect of bases on litmus
- describe the effects of acid on blue litmus
- write example equations illustrating the reaction of acids and bases with water
- write the formula for the hydronium ion
- give example equations illustrating the dissolving of a base in water
- describe what is meant by a salt and how they can be formed
- write example equation illustrating the neutralization of an acid with base
- tell what an acidic solution is in terms of relative concentrations of OH^{1-} and H_3O^{1+}

Key terms

hydronium ion, H_3O^{1+}	litmus	ionization
base, proton acceptor	acid, proton donor	hydroxide ion, OH^{1-}
acidic solution	basic solution	neutralization reaction
acid-base reaction	salt	equivalent amounts of acid & base
ammonia	ammonium	completion of reaction
water-soluble compound	alkaline solution	acid-base indicators

7.2 The Strengths of Acids and Bases

The main ideas in this section are the descriptions of equilibrium, extent of ionization, strong acids, weak acids, strong bases, weak bases, strong electrolytes, weak electrolytes.

Objectives
After studying this section a student should be able to:
- explain what is meant by extent of ionization
- tell what is meant by slightly ionized
- describe what is meant by a strong acid and a strong base in terms of % ionization
- describe what is meant by a weak acid and a weak base in terms of % ionization

Key terms

extent of ionization	weak acid	slightly ionized
strong acid	100% ionized	strong base
strong electrolyte	weak base	weak electrolyte
equilibrium		

7.3 Molarity and the pH Scale

This section describes the self ionization of water, the ion product for water, molarity as a concentration measure. The calculations involving molarity, grams of solute, molar mass, and volume of solution are illustrated. Acidity is linked to the definition of pH, where $pH = -\log [H_3O^{1+}]$. Examples are given for using the pH scale range and how to determine pH from H_3O^{1+} and OH^{1-}. The pH values for common substances are described.

Objectives
After studying this section a student should be able to:
- tell what is meant by concentration
- describe what is meant by a neutral solution
- state the value for the ion product $[H_3O^{1+}][OH^{1-}]$ for water
- describe how $[H_3O^{1+}]$ and $[OH^{1-}]$ in water are controlled by the constant 1×10^{-14}
- calculate $[H_3O^{1+}]$ from $[OH^{1-}]$ and the ion product for water, $[H_3O^{1+}][OH^{1-}] = 1 \times 10^{-14}$
- calculate $[OH^{1-}]$ from $[H_3O^{1+}]$ and the ion product for water, $[H_3O^{1+}][OH^{1-}] = 1 \times 10^{-14}$
- give the definition for pH and tell the pH ranges for acidic, basic and neutral solutions
- tell whether a solution is acidic, basic or neutral given the pH
- calculate pH for a solution from a known $[H_3O^{1+}]$
- calculate pH for a solution from $[OH^{1-}]$ and the relation $[H_3O^{1+}][OH^{1-}] = 1 \times 10^{-14}$
- determine $[H_3O^{1+}]$ from pH using the relation $[H_3O^{1+}] = 10^{-pH}$
- tell how a strong concentration and a strong acid differ
- use $M = \dfrac{\text{\# moles solute}}{\text{\# liters of solution}}$ to calculate molarity from raw data and to calculate moles of solute needed to prepare a solution of specific molarity and volume
- calculate grams of solute needed to prepare a solution of specific molarity and volume

Key terms

neutral
acidic pH range
$M = \dfrac{\text{\# moles solute}}{\text{\# liters of solution}}$
$[H_3O^{1+}] = 10^{-pH}$

ion product
pH
$pH = -\log [H_3O^{1+}]$

\# grams solute = \# moles x molar mass

molarity, M and [ ]
basic pH range
$[H_3O^{1+}][OH^{1-}] = 1 \times 10^{-14}$

7.4 Acid-Base Buffers

This section gives the definition of a buffer, explains how buffers are made and how they work to stabilize pH. The role of acid-base buffers in the human body is introduced. The carbonic acid-bicarbonate ion, H_2CO_3 / HCO_3^{1-}, and the acetic acid-acetate ion, CH_3COOH/CH_3COO^{1-} buffer systems are described.

Objectives
After studying this section a student should be able to:
- tell what buffer solutions are
- tell what combinations of compounds are required to make a buffer solution
- tell what is the normal range for blood pH
- name the two conditions that occur when blood pH goes outside the normal range
- describe how a buffer system like H_2CO_3 / HCO_3^{1-} regulates pH
- write equations showing how a buffer system like H_2CO_3 / HCO_3^{1-} regulates pH when either acid or base is added
- write equations showing how a buffer controls pH when hydronium ion is added

Key terms
- acid-base buffer
- alkalosis
- acidosis
- blood pH
- H_2CO_3 / HCO_3^{1-} buffer system
- CH_3COOH/CH_3COO^{1-} buffer system

7.5 Corrosive Cleaners

This section discusses both acidic and basic cleaners with extreme values in pH. The applications and properties of strong alkaline cleaners like sodium hydroxide (NaOH) are described in detail. Safe practices in using these products are explained.

Objectives

After studying this section a student should be able to:
- describe the proper way to handle corrosive cleaners
- name the most common strong alkali used in drain cleaners and oven cleaners
- explain why drain cleaners often contain both aluminum hydroxide and sodium hydroxide
- tell why aerosol spray oven cleaners must be used carefully

Key terms
- corrosive cleaners
- oven cleaners
- drain cleaners
- toilet-bowl cleaners
- hydrochloric acid, HCl
- caustics
- phosphoric acid, H_3PO_4 (aq)
- oxalic acid, $H_2C_2O_4$ (aq)
- sodium hydroxide, NaOH

7.6 Heartburn: Why Reach for an Antacid

The normal pH level of human stomach contents is discussed. Compounds in commercial antacids are described. The effect of antacid tablets on stomach acid is illustrated with reactions.

Objectives

After studying this section a student should be able to:
- tell what is normal stomach pH
- tell what acid is secreted by the stomach lining
- describe the connection between "heartburn" and stomach pH
- name two common antacids and their active ingredients.

Key terms
- stomach acid pH = (1.5-2.0)
- baking soda
- heartburn
- bicarbonate, HCO_3^{1-}
- antacid

Additional readings

Cramer, Tom. "What's in an antacid?" FDA Consumer v 26, Jan-Feb 1992: 21.

Davenport, Horace W. "Why The Stomach Does Not Digest Itself." Scientific American Jan. 1972 : 86 .

"Drain Cleaners: Dealing with your drains." Consumer Reports Jan. 1994: 44-48.

Husar, John. " Catch on to the pH factor; better catches may follow." Chicago Tribune 22 October 1989, Sports.

Mebane, R.C., & Reybold, T. "Edible acid-base indicators." Journal of Chemical Education, 62, (1985): 285.

McNulty, Karen. "Biscuit blues. (acid-base chemistry in the kitchen)." <u>Science World</u> 11 Mar. 1994: 18.

Answers to Odd Numbered Questions for Review and Thought

1. a. Acidic solutions have H_3O^{1+} concentrations that are greater than the OH^{1-} concentration. The pH is less than 7 for acidic solutions.
    b. Basic solutions have OH^{1-} concentrations that are greater than the H_3O^{1+} concentration. Basic solutions have a pH greater than 7.
    c. Neutral solutions have pH = 7 and have equal concentrations of hydronium ions, H_3O^{1+}, and hydroxide ions, OH^{1-}.

3. A salt is the substance formed between the anion of an acid and the cation of a base. An example is KBr, a typical salt formed in the reaction shown here.
    $$KOH(aq) + HBr(aq) \longrightarrow KBr(aq) + H_2O(\ell)$$

5. a. The extent of ionization measures the fraction of molecules that ionize.
    b. A strong acid is dissociates or ionizes 100% to form and an anion.
    $$HI(aq) + H_2O(\ell) \longrightarrow H_3O^{1+}(aq) + I^{1-}(aq)$$
    c. A strong base is a substance that is essentially completely ionized.
    d. A weak acid is only partially ionized so that the majority of acid molecules or ions are still intact.
    e. A weak base is only partially ionized so that the majority of base molecules or ions are intact.

7. Common basic substances are soap, baking soda, household ammonia and milk of magnesia.

9. a. Gastric fluid, acidic   b. Tomatoes, acidic
    c. Oven cleaner, basic   d. Soap, basic
    e. Carbonated beverages, acidic   f. Baking soda, basic

11. a. An acid-base indicator changes color with changes in acidity or pH. The indicator molecules act as both acid and base. They display one color when protonated in acidic solutions and another color when they lose a proton in basic solutions. Litmus is blue in basic solutions and turns red in acidic solutions. It is sometimes called "red litmus" when it is red and "blue litmus" when it is blue.
    b. Molarity is a concentration measure equal to the number of moles of solute in one liter of solution; it is calculated by dividing the # moles of solute by the # of liters of solution.
    c. Concentration measures the relative amount of dissolved solute in a definite amount of solution.

13. An acidic buffer is an aqueous mixture of both a weak acid and its anion that will maintain a stable pH when either acid or base is added. The acid will neutralize added base and the anion will neutralize added acid. A basic buffer is a mixture of a weak base and its cation. It will also maintain a stable pH because the weak base will neutralize additional acid and the cation will react with added base.

15. a. $CH_3COOH_{(aq)} + KOH_{(aq)} \longrightarrow H_2O_{(\ell)} + CH_3COOK_{(aq)}$
    b. $H_2SO_{4(aq)} + Ca(OH)_{2(aq)} \longrightarrow CaSO_{4(aq)} + 2\,H_2O_{(\ell)}$
    c. $H_2SO_{4(aq)} + 2\,NaOH_{(aq)} \longrightarrow Na_2SO_{4(aq)} + 2\,H_2O_{(\ell)}$

17. The total molecular equation for the reaction of stomach acid and Tums® is
    $2\,HCl_{(aq)} + CaCO_{3(s)} \longrightarrow CaCl_{2(s)} + CO_{2(g)} + H_2O_{(\ell)}$.
    Stomach acid is an aqueous solution of hydrochloric acid, HCl(aq). The active ingredient in Tums® is $CaCO_{3(s)}$.

19. The black coffee with the pH = 5.0 is more acidic than the milk with a pH of 6.5. Solutions with pH = 7 are neutral. Solutions with pH below 7 are acidic and those with pH above 7 are basic. The lower the pH the more acid the solution. The higher the pH the less acidic the solution.

21. Gastric juice is more acidic than tomato juice. Gastric juice has a pH = 1 and gives a $[H_3O^{1+}] = 10^{-pH}$ so $[H_3O^{1+}] = 10^{-1} = 0.10$. Tomato juice has a pH = 4 and gives a $[H_3O^{1+}] = 10^{-pH}$ so $[H_3O^{1+}] = 10^{-4} = 0.0001$.
    The relative acidity of gastric juice to tomato juice can be figured by dividing the gastric juice H_3O^{1+} concentration by the tomato juice H_3O^{1+} concentration.

    $$\frac{\text{Gastric juice }[H_3O^+]}{\text{Tomato juice }[H_3O^+]} = \frac{10^{-1}\,\text{moles H}^{1+}/\text{liter}}{10^{-4}\,\text{moles H}^{1+}/\text{liter}} = 10^{-1} \times 10^4 = 10^3 = 1000$$

    Gastric juice is 1000 times more acidic than tomato juice.

23. a. Generally NaOH(aq) and HCl(aq) would not be a good combination for a buffer because a neutralization reaction would occur between the acid and base. If excess HCl were mixed with NaOH then the base would be completely neutralized. If excess NaOH were mixed with a smaller amount of HCl then the acid would all be neutralized. In both cases the mixture would not contain one component needed to create a buffer.
    b. A mixture of CH_3COOH and CH_3COONa would be a good combination to prepare a buffer. Any added base would be neutralized by the acidic CH_3COOH. Any added acid would be neutralized by the slightly basic anion, CH_3COO^{1-}.

25. A stronger acid gives more H_3O^{1+} ions for the same acid concentration. The stronger acid gives more H_3O^{1+} ions because it is essentially 100% dissociated and all of the acid molecules are converted to ions. The pH will be lower for the strong acid than for the weak acid. The pH = 4.6 solution of A will be the weak acid solution and the more acidic solution of B with pH = 1.1 will be the stronger acid.

Solutions for Problems

1. The molarity is determined from the number of moles of solute and the volume of solution in liters. The grams of solute must be converted to moles. The molar mass is needed for this conversion of grams to moles. The volume of solution must be converted to liters if it is in any other volume unit. The answer is rounded to the correct number of significant digits.

   a. The 5.0 g HCl must be converted to moles of HCl.
   Determine the molar mass for the solute by adding the mass each element contributes.
   Molar mass for HCl = 1.01 g H + 35.45 g Cl = 36.46 g HCl
   The volume of solution is 1.0 liters.
   Determine the molarity.
   $$\text{molarity HCl} = \frac{\text{moles solute HCl}}{\text{Liters solution}} =$$
   Calculate the number of moles of solute
   $$\text{moles HCl} = 5.0 \text{ g HCl} \times \frac{1 \text{ mole HCl}}{36.46 \text{ g HCl}} = 0.14 \text{ moles HCl}$$
   Divide the number of moles by the number of liters.
   $$\text{molarity HCl} = \frac{\text{moles solute HCl}}{\text{Liters solution}} = \frac{0.14 \text{ moles HCl}}{1.0 \text{ Liters solution}} = 0.14 \text{ M HCl}$$
   The calculation can be done in one set up:
   $$5.0 \text{ g HCl} \times \frac{1 \text{ mole HCl}}{36.46 \text{ g HCl}} \times \frac{1}{1.0 \text{ L}} = 0.14 \text{ M HCl}$$

   b. First convert the 40.0 g NaOH to moles of NaOH.
   Molar mass for NaOH = 1.01 g H + 15.99 g O + 22.99 g Na = 39.99 g NaOH
   The molar mass rounded to three significant digits is 40.0g NaOH.
   The 1.0 L of solution is already in units of liters.
   $$\text{Molarity NaOH} = 40.0 \text{ g NaOH} \times \frac{1 \text{ mole NaOH}}{40.0 \text{ g NaOH}} \times \frac{1}{1.0 \text{ L}} = 1.0 \text{ M NaOH}$$

   c. Determine the molar mass for the NaCl.
   Molar mass for NaCl = 22.99 g Na + 35.45 g H = 58.44 g NaCl.
   $$\text{Molarity NaCl} = \frac{58.5 \text{ g NaCl}}{1} \times \frac{1 \text{ mole NaCl}}{58.44 \text{ g NaCl}} \times \frac{1}{1.0 \text{ L}} = 1.00 \text{ M NaCl}$$
   The answer is rounded off to two significant digits 1.0 M

   d. Determine the molar mass for the solute by adding the mass each element contributes.
   Molar mass for NaCl = 22.99 g Na + 35.45 g H = 58.44 g NaCl.
   $$\text{Molarity NaCl} = \frac{1.0 \text{ g NaCl}}{1} \times \frac{1 \text{ mole NaCl}}{58.44 \text{ g NaCl}} \times \frac{1}{1.0 \text{ L}} = 0.0171 = 0.017 \text{ M NaCl}$$
   Round answer to the correct number of significant digits, s.d. Relationships within the metric system are exact, 1 L = 1000 mL, and do not affect the number of s.d. in an answer.

2. The number of grams of solute in a solution can be calculated using the molarity and the volume in liters as well as the molar mass of the solute. The number of moles is determined from the molarity and the volume of the solution.
   moles solute = M x V = (molarity) (volume of solution in liters)
   Then, g solute is calculated from g solute = moles x molar mass of solute.
   The number of grams of solute can be calculated by multiplying number of moles by the molar mass.

   Determine the molar mass for the solute by adding the mass each element contributes.
   Molar mass for NaCl = 22.99 g Na + 35.45 g H = 58.44 g NaCl
   The volume of solution is in liters so no conversion is needed.
   Use the molarity and volume to determine the number of moles of solute.
   moles NaCl = M x V = (1.5 mole NaCl / L) x (1.0 L) = 1.5 moles NaCl
   Calculate the number of grams of solute using the moles of solute and the molar mass:
   g NaCl = moles solute x molar mass
   $$g\ NaCl = 1.5\ moles NaCl \times \frac{58.44g\ NaCl}{1\ mole\ NaCl} = 87.66\ g\ NaCl$$
   The calculation can be done in one set up:
   $$g\ NaOH = \frac{1.5\ mole\ NaCl}{1\ liter} \times \frac{1.0\ L}{1} \times \frac{58.44\ g\ NaCl}{1\ mole} = 87.66\ g\ NaCl = 88.\ g\ NaCl$$
   The 87.66 g NaCl is rounded off to 88. g because the 1.5 M NaCl and the 1.0 L have two significant digits each.

3. The number of grams of solute in a solution can be calculated using the molarity and the volume in liters as well as the molar mass of the solute. The number of moles is determined from the molarity and the volume of the solution.
   moles solute = M x V = (molarity) (volume of solution in liters)
   Then the g solute is calculated from g solute = moles x molar mass of solute.
   An alternative way to solve this problem is to first determine the moles of solute in one mL of solution. Calculate the total moles of solute by multiplying the moles in one mL by the number of mLs.
   The number of grams of solute can be calculated by multiplying number of moles by the molar mass. The formula for sulfuric acid is H_2SO_4.
   Molar mass H_2SO_4 = 2x1.01 g H + 1x32.06 g S + 4x16.00 g O = 98.08 g H_2SO_4
   $$g\ H_2SO_4 = \frac{0.10\ mole\ H_2SO_4}{1\ Liter} \times \frac{1\ Liter}{1} \times \frac{98.08\ g\ H_2SO_4}{1\ mole\ H_2SO_4} = 9.808\ g\ H_2SO_4$$
   The answer rounded off to two significant digits is 9.8 g H_2SO_4.

4. The pH for a solution is determined from the definition pH = - log[H_3O^{1+}]. A simple graphic relationship between the hydronium ion concentration and pH is shown here. The top row gives the concentration [H_3O^{1+}] and the second line gives the matching pH. This is limited only to the whole numbered exponents of 10.

[H_3O^{1+}]	10^0	10^{-1}	10^{-2}	10^{-3}	10^{-4}	10^{-5}	10^{-6}	10^{-7}	10^{-8}	10^{-9}	10^{-10}	10^{-11}	10^{-12}	10^{-13}	10^{-14}
pH	0	1	2	3	4	5	6	7	8	9	10	11	12	13	14

a. Hydrochloric acid, HCl(aq) is a strong acid. The $[H_3O^{1+}] = 1.0 \times 10^{-2}$ because strong acids are 100% ionized and the ion concentrations equal the original acid concentration.
$HCl_{(aq)} + H_2O_{(\ell)} \longrightarrow H_3O^{1+}{}_{(aq)} + Cl^{1-}{}_{(aq)}$
$pH = -\log[H_3O^{1+}]$; $pH = -\log 1 \times 10^{-2}$;
The log function on a calculator gives pH = 2
or the graphical relationship shown above can be used to get the same result.

b. Nitric acid, $HNO_{3(aq)}$, is a strong acid and strong acids are 100% ionized.
$HNO_{3(aq)} + H_2O_{(\ell)} \longrightarrow H_3O^{1+}{}_{(aq)} + NO_3^{1-}{}_{(aq)}$
so the $[H_3O^{1+}] = [NO_3^{1-}] = 1.0 \times 10^{-3}$
$pH = -\log[H_3O^{1+}]$; $pH = -\log 1 \times 10^{-3}$; pH = 3

c. Sodium hydroxide is a strong base. A strong base is completely ionized. The Na^{1+} and OH^{1-} concentrations equal the original NaOH concentration. $[Na^+] = [OH^{1-}] = 0.1 = 1 \times 10^{-1}$. Use the relationship $Kw = [H_3O^{1+}][OH^{1-}] = 1 \times 10^{-14}$ to calculate the $[H_3O^{1+}]$. Substitute 1×10^{-1} for the hydroxide concentration; $1 \times 10^{-14} = [H_3O^{1+}][OH^{1-}] = [H_3O^{1+}][1 \times 10^{-1}]$;
$[H_3O^{1+}] = \dfrac{1 \times 10^{-14}}{[1 \times 10^{-1}]} = \dfrac{1}{1} \times 10^{-14--1} = 1 \times 10^{-14+1} = 1 \times 10^{-13}$. When the coefficients are 1, a simpler method recognizes that the sum of the exponents on ten on both sides must equal -14 because $[H_3O^{1+}][OH^{1-}] = 1 \times 10^{-14}$. This means -1 + ? = -14; The unknown exponent must be -13. so $[H_3O^{1+}] = 1 \times 10^{-13}$.
$pH = -\log[H_3O^{1+}]$; $pH = -\log 1 \times 10^{-13}$; pH = 13.

d. Hydrobromic acid, HBr, ionizes 100%. The concentration of hydronium ion, $[H_3O^{1+}]$, is 0.10 M = 1.0×10^{-1}.
Using $pH = -\log[H_3O^{1+}]$; $pH = -\log 1 \times 10^{-1}$; pH = 1

5. The pH for a solution is determined from the definition $pH = -\log[H_3O^{1+}]$. A simple graphic relationship between the hydronium ion concentration and pH is shown here. The top row gives the hydroxide ion concentration, $[OH^{1-}]$. The second row gives the hydronium ion concentration, $[H_3O^{1+}]$. The third line gives the matching pH. For simplicity this graphic is limited to concentrations that are whole numbered powers of 10, that is $1 \times 10^{-\#}$M.

$[OH^{1-}]$	10^{-14}	10^{-13}	10^{-12}	10^{-11}	10^{-10}	10^{-9}	10^{-8}	10^{-7}	10^{-6}	10^{-5}	10^{-4}	10^{-3}	10^{-2}	10^{-1}	10^{0}
$[H_3O^{1+}]$	10^{0}	10^{-1}	10^{-2}	10^{-3}	10^{-4}	10^{-5}	10^{-6}	10^{-7}	10^{-8}	10^{-9}	10^{-10}	10^{-11}	10^{-12}	10^{-13}	10^{-14}
pH	0	1	2	3	4	5	6	7	8	9	10	11	12	13	14

a. Sodium hydroxide is a strong base. The Na^{1+} and OH^{1-} concentrations equal the original NaOH concentration. $[Na^+] = [OH^{1-}] = 0.001 = 1 \times 10^{-3}$.
    Use the relationship $K_w = [H_3O^{1+}][OH^{1-}] = 1 \times 10^{-14}$ to calculate the $[H_3O^{1+}]$.
    Substitute 1×10^{-3} for the hydroxide concentration;
    $$1 \times 10^{-14} = [H_3O^{1+}][OH^{1-}] = [H_3O^{1+}][1 \times 10^{-3}] ;$$
    Solve for $[H_3O^{1+}]$
    $$[H_3O^{1+}] = \frac{1 \times 10^{-14}}{[1 \times 10^{-3}]} = \frac{1}{1} \times 10^{-14-3} = 1 \times 10^{-14+3} = 1 \times 10^{-11}.$$
    When the coefficients are 1, a simpler method recognizes that the sum of the exponents on ten on both sides must equal -14 because $[H_3O^{1+}][OH^{1-}] = 1 \times 10^{-14}$. This means -3 + ? = -14; The unknown exponent must be -11. so $[H_3O^{1+}] = 1 \times 10^{-11}$.
    $pH = -\log[H_3O^{1+}]$; $pH = -\log 1 \times 10^{-11}$; $pH = 11$.

b. Hydrochloric acid, HCl(aq) is a strong acid. The $[H_3O^{1+}] = 1.0 \times 10^{-3}$ because strong acids are 100% ionized $HCl_{(aq)} + H_2O_{(\ell)} \longrightarrow H_3O^{1+}_{(aq)} + Cl^{1-}_{(aq)}$
    The ion concentrations equal the original acid concentration.
    $pH = -\log[H_3O^{1+}]$; $pH = -\log[1.0 \times 10^{-3}] = -\log 0.001 = 3$;
    The log function on a calculator gives pH = 3
    or the graphical relationship shown above can be used to get the same result.

c. Potassium hydroxide is a strong base. The K^{1+} and OH^{1-} concentrations equal the original KOH concentration. $[K^+] = [OH^{1-}] = 0.01 = 1 \times 10^{-2}$.
    Use the relationship $K_w = [H_3O^{1+}][OH^{1-}] = 1 \times 10^{-14}$ to calculate the $[H_3O^{1+}]$.
    Substitute 1×10^{-2} for the hydroxide concentration;
    $$1 \times 10^{-14} = [H_3O^{1+}][OH^{1-}] = [H_3O^{1+}][1 \times 10^{-2}] ;$$
    Solve for $[H_3O^{1+}]$
    $$[H_3O^{1+}] = \frac{1 \times 10^{-14}}{[1 \times 10^{-2}]} = \frac{1}{1} \times 10^{-14-2} = 1 \times 10^{-14+2} = 1 \times 10^{-12}.$$
    When the coefficients are 1, a simpler method recognizes that the sum of the exponents on ten on both sides must equal -14 because $[H_3O^{1+}][OH^{1-}] = 1 \times 10^{-14}$. This means -2 + ? = -14; The unknown exponent must be -12. so $[H_3O^{1+}] = 1 \times 10^{-12}$.
    $pH = -\log[H_3O^{1+}]$; $pH = -\log 1 \times 10^{-12}$; $pH = 12$.

d. Neutral water has equal concentrations for $[H_3O^{1+}]$ and $[OH^{1-}]$. The product of these concentrations equals the K_w.
    $K_w = [H_3O^{1+}][OH^{1-}] = 1 \times 10^{-14}$; Let $x = [H_3O^{1+}] = [OH^{1-}]$
    $[x][x] = 1 \times 10^{-14}$; $x^2 = 1 \times 10^{-14}$; $x = \sqrt{1 \times 10^{-14}} = \sqrt{1} \times 10^{-14 \div 2} = 1 \times 10^{-7}$;
    $pH = -\log[H_3O^{1+}]$; $pH = -\log 1 \times 10^{-7}$; $pH = 7$

6. The definition of pH is pH = -log[H_3O^{1+}]. The inverse operation of taking the logarithm is to raise 10 to a power equal to the logarithm. This means molarity = [H_3O^{1+}] = 10^{-pH}

   a. pH = 1 ; [H_3O^{1+}] = 10^{-pH};   [H_3O^{1+}] = 10^{-1} = 1 x 10^{-1} or 0.1
   b. pH = 0 ; [H_3O^{1+}] = 10^{-pH};   [H_3O^{1+}] = 10^{-0} = 1 x 10^{-0} or 1
   c. pH = 5 ; [H_3O^{1+}] = 10^{-pH};   [H_3O^{1+}] = 10^{-5} = 1 x 10^{-5} or 0.00001
   d. pH = 3 ; [H_3O^{1+}] = 10^{-pH};   [H_3O^{1+}] = 10^{-3} = 1 x 10^{-3} or 0.001

Speaking of Chemistry

Name _____

Acid–Base Reactions

Across
2. Indicators change____ when pH changes
6. Name for H_3O^{1+} ion
9. Dyes that change color when pH changes
10. Blue litmus turns_____
11. Positive charge on hydronium ion
14. Process of reacting an acid with a base
18. Opposite of an acid
19. Neutral solution pH
22. Number of ions formed when KOH dissolves in water
23. Body fluid buffered to pH =7.35

Down
1. Acids are _____ donors.
2. NaOH is a _____ cleaner
3. The pH for $[H_3O^{1+}] = 1 \times 10^{-1}$.
4. Acids have a _____ taste.
5. Solution that has stable pH when either acid or base is added
6. Name for OH^{1-} ion
7. A person can _____ from acidosis slippery basic cleaning substance.
8. Acetic acid is a _____ acid.
12. Sulfuric acid is a _____ acid
13. Concentration measure with symbol, M.
15. $[H_3O^{1+}] = 1 \times 10^{-11}$ has pH = ___.
16. Describes proton donors
17. Tums is an _____
20. The pH when $[OH^{1-}] = 1 \times 10^{-10}$
21. Formula for potassium hydroxide

© 1999 Harcourt Brace & Company. All rights reserved.

Bridging the Gap   I
Acidity, pH, Indicator Paper and Red Cabbage

Acids and bases are part of us and our surroundings. The body uses hydrochloric acid, HCl(aq) in the stomach during digestion. Calcium carbonate, $CaCO_3(s)$, is the active ingredient in Tums™ an antacid used to treat excess stomach acid. These substances all yield solutions with characteristic pH values. Several pH sensitive dyes can be used to estimate pH values for solutions. You will use these as your pH indicator.

Red cabbage contains natural dyes that change color with acidity changes. The color displayed by these dyes is a good indicator of the pH. The color of these dyes is shown in the text on page 168.

This plot is a graphical summary of the relationship between color and pH. Acid solutions will have a red or pink color. Neutral solutions will have a pretty lavender color and basic solutions will have a blue-green or yellow color.

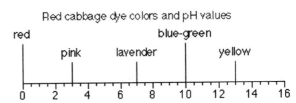

Red cabbage indicator solution

You will need a few leaves of red cabbage, a blender or food processor, a coffee filter or food strainer, and a clean container to collect the cabbage juice. Tear one or two cabbage leaves into pieces and place these in the blender with about 2 cups of water; blend for one or two minutes, then filter the mixture through the coffee filter. Collect and save the red cabbage juice.

1. Test at least three different household chemicals such as baking soda, vinegar, household ammonia, soap, milk, 7-Up™ or club soda; each must be dissolved in water. Place each solution in a clear, colorless container such as a glass or jelly jar.
2. Place a piece of white paper on your work surface. This gives a good background for seeing the solution colors. Label the different containers.
3. Add about 10 mL, 2 teaspoons, of cabbage extract to each solution and stir; note the color of your liquid. Record the results of the different tests in the table on the report sheet.

Indicator paper

1. Take a 2-inch-wide strip of white facial tissue and tape the upper end to a pencil so the tissue hangs down and can be lowered into the cabbage indicator solution. Put about 1/3 cup of cabbage dye into a quart jar or other tall container. Lower the strip of paper into the jar until its lower end is in the cabbage dye. Rest the pencil across the top of the container so the paper hangs inside. Let the solution wick up the paper until the liquid has climbed to about 6 inches. This will take about 40 minutes.
2. Remove the wet paper from the liquid and hang it up to dry.
3. When the strip of facial tissue is thoroughly dry, cut it into pieces about 1 inch x 1 inch. Store them in a brown jar or dark place.
4. This paper will act as an acid base indicator. It will turn pink when touched with a drop of acid and turn blue-green when touched with a drop of base; try it. Save the indicator paper for your own tests.

© 1999 Harcourt Brace & Company. All rights reserved.

Bridging the Gap I Name _____
Acidity, pH, Indicator Paper and Red Cabbage

Data and Observations

Household substance	Cabbage juice indicator color	Estimated pH	pH using page 266

Analysis and Conclusions

Do you believe the estimated experimental pH values and the table values agree? Remember these are only estimates and are not exact. Justify your answer.

Bridging the Gap II
Acid Concentration and Red Cabbage Dyes

A common base is household ammonia which is a mixture of water and ammonia, $NH_{3(g)}$. The strong, biting odor is from the dissolved $NH_{3(g)}$ that escapes from the solution. Household ammonia that has been open for a time will not be as strong as a "fresh" solution. Vinegar is an acid solution which contains acetic acid, $HC_2H_3O_2$, and water. In this exercise you will determine the relative strength of your vinegar and compare the result with the concentration listed in the text. Natural red cabbage juice indicator is used to tell when the solution pH is neutral and the ammonia has just neutralized the vinegar.

Equipment and materials
Five clear colorless glass or acrylic tumblers that can hold 1 cup of liquid, plastic straws, white paper, clear uncolored household ammonia, white vinegar, distilled water or tap water.

1. Place the piece of white paper on your work surface. This gives a good background for seeing the solution colors. Mark the positions for the different containers.
2. Place one glass in the middle. Place a glass on either side of the first one.
3. Pour approximately 30 mL (2 tablespoons) of vinegar into the glass on the left. Rinse your tablespoon with tap water. Add a similar amount of household ammonia to the glass on the right. Place a plastic straw in each glass.
4. Add two teaspoons of cabbage juice to the middle glass. Add 1/2 cup of water to this glass of cabbage juice. Check to see what the color is for this water mixture.
5. Add approximately 30 mL of household ammonia and 2 teaspoons of cabbage dye to the 4th glass; mix and set aside to the far right as a reference for the base color.
6. Place approximately 30 mL of vinegar and 2 teaspoons of cabbage dye in the 5th glass; mix and set aside to the far left as a reference for the acid color.
7. Use the straw to transfer <u>exactly 20 drops</u> of plain vinegar to the middle glass that already contains the water and cabbage juice mixture. Do not draw any liquid into the straw using your mouth. The straw will collect some liquid inside because it is open and there will be some capillary rise. Cap off the top of the straw with a finger. Some liquid will stay in the straw.

Carefully move the straw over to the middle glass and squeeze the straw to deliver individual drops of vinegar. You will have to do a series of transfers to get 20 drops. Note the color.

8. Now transfer ammonia solution <u>one drop at a time</u> to the middle glass containing the mixture of water, cabbage dye and 20 drops of vinegar. Swirl or stir and note the color after each drop. Count the number of drops of ammonia needed to produce the lavender color of a neutral solution.
9. Record the number of drops of ammonia solution used to neutralize 20 drops of vinegar. Based on your work, which of the two solutions has the higher concentration? Use the ratio of # drops of ammonia to # drops of vinegar to find the molarity of the vinegar.
10. Wash out the middle glass, rinse with tap water; repeat Steps 3, 7, 8, and 9.

Bridging the Gap   II   Name _____
Acid Concentration using Red Cabbage Dyes

Data

	Trial 1	Trial 2
Number of drops of vinegar	20 drops	20 drops
Number of drops of ammonia		

Calculations

Household ammonia is 2.5 M NH$_3$. Calculate the molarity of your vinegar using the equation below and record your result for each trial. Then average your vinegar molarity values.

$$M_{vinegar} = 2.5 \times \frac{\text{\# drops ammonia}}{\text{\# drops vinegar}} = \text{molarity of your vinegar}$$

Trial 1

$$M_{vinegar} = 2.5 \times \frac{\text{drops ammonia}}{\text{drops vinegar}}$$

$M_{vinegar} =$

Trial 2

$$M_{vinegar} = 2.5 \times \frac{\text{drops ammonia}}{\text{drops vinegar}}$$

$M_{vinegar} =$

Average $M_{vinegar} =$

Analysis and Conclusions

1. The usual molarity for acetic acid in vinegar is about 0.7 M. Do you feel your experimental value agrees with this

2. Name another acidic solution you could analyze with this neutralization method.

© 1999 Harcourt Brace & Company.   All rights reserved.

Bridging the Gap   III
Internet access to the United States Geological Survey, USGS
National Water Conditions, pH of Precipitation, Acid Rain Where You Live

The USGS : National Water Conditions

The United States Geological Survey is a part of the Department of the Interior. Specific activities of the USGS have separate Internet home pages. The monitoring of the National Water Condition is one of the responsibilities of the USGS. The "Table of Contents" page for the various activities is readily accessible using the following uniform resource locator (URL) web site address.

http://h2o.usgs.gov/nwc/NWC/html/TOC.html

pH of Precipitation: map of the United States

You can scroll down the table of contents page and click on the active line named

"•pH of Precipitation".

This will give you the "pH of Precipitation" page. You can go to this page directly if you use the following URL

http://h2o.usgs.gov/nwc/NWC/pH/html/ph.html

This page gives a map of the United States with state boundaries. The pH values are updated every month. Data from earlier reports can be also accessed. There are 190 monitoring sites on the map. The point and click feature of the map lets you choose a state and get rainfall pH values from the sites in that state. You can either print the image or save it to disk.

The purpose of this assignment is to open the "pH of Precipitation" page under "National Water Conditions" and select a state of your choice. You are to record the name of the state, sketch its outline (alternately you can download the map). You are to record the pH readings in that state.

You are also to note the pH values for the states east of the Mississippi River. You are to find the state in this group with the lowest value pH for its precipitation and record both the pH and the identity of the state.

You are supposed to record the values for pH of rainfall in Oregon and the pH for points in Ohio. If there are differences, you are supposed to give some explanation for them.

© 1999 Harcourt Brace & Company. All rights reserved.

Bridging the Gap   III   Name _____
National Water Conditions, pH of Precipitation, Acid Rain Where You Live

1. Give the uniform resource locator (URL) for the USGS "pH of Precipitation" page that shows a map of the United States and pH values across the country.

    http://_____

    What state east of the Mississippi has the most acidic rainfall? (the lowest pH)

2. Identify the state you chose to check. What are the pH values for sites in this state? Are the reported values more acid than normal rainfall with a pH of 5.6?

3. Give the uniform resource locator (URL) for the state you selected.

    http://_____

4. What is the pH for rainfall in Oregon?

    What is the pH for rainfall in Ohio?

    Which of these is more acidic?

    What environmental reasons exist for the differences?

8 Oxidation-Reduction Reactions

8.1 Oxidation and Reduction

This section describes oxides as compounds like Al_2O_3 and SO_3. Reduction is identified as a gain of electrons such as $Cl_2(g) + 2\ e^- \longrightarrow 2\ Cl^{1-}$. Oxidation is defined as a loss of electrons such as $Al(s) \longrightarrow Al^{3+} + 3\ e^-$. Example reactions are used to illustrate each definition. A substance that has combined with oxygen is described as being oxidized. All combustion reactions are oxidation-reduction reactions. For example the carbon atom is oxidized and the oxygen atoms are reduced in the reaction $CH_4(g) + 2\ O_2(g) \longrightarrow CO_2(g) + 2\ H_2O(g)$; the hydrogen atoms are not involved in either the oxidation or reduction. Rusting is illustrated and shown to be an oxidation-reduction (redox) reaction. Redox reactions always require both oxidation and reduction to simultaneously occur. Substances that are oxidized are identified in equations. Substances that are reduced are identified in equations.

Objectives
After studying this section a student should be able to:
- give the definition for an oxide and give formulas and names for common oxides
- state the definition for reduction
- inspect a reaction and tell whether or not reduction occurred
- inspect a reaction and identify the reducing agent and the oxidizing agent
- give a definition and example for combustion
- state the definition for oxidation and give an example
- identify the oxidized substance and the reduced substance in a reaction
- inspect a reaction equation and tell whether or not oxidation occurred

Key terms
reduction	gain of hydrogen	oxidation-reduction reaction
oxidation	oxidized	loss of electrons
combustion	redox reactions	oxide

8.2 Oxidizing Agents: They Bleach and They Disinfect

This section describes oxidizing agents and their reactions. Examples like bleach, NaOCl, and hydrogen peroxide, H_2O_2, are listed. Their reactions are illustrated. The action of bleach to cause whitening of stains is explained.

Objectives
After studying this section a student should be able to:
- give the definition for an oxidizing agent
- identify the oxidizing agent in an equation
- give some examples and uses of oxidizing agents like H_2O_2

Key terms
oxidizing agent	hydrogen peroxide, H_2O_2	oxidation
halogens, $F_2, Cl_2, Br_2, ...$	disinfectant	bleach, NaOCl

8.3 Reducing Agents for Metallurgy and Good Health

Reducing agents are defined as substances that lose hydrogen, lose electrons and gain oxygen like carbon in the combustion of methane. $CH_4(g) + 2\ O_2(g) \longrightarrow CO_2(g) + 2\ H_2O(g)$ Copper is oxidized and is a reducing agent in the reaction $2\ Cu(s) + O_2(g) \longrightarrow 2\ CuO(s)$

© 1999 Harcourt Brace & Company. All rights reserved.

Objectives
After studying this section a student should be able to:
- give the definition for an oxidizing agent and give applications for oxidizing agents like F_2
- state definitions for reducing agent
- give some examples and uses of reducing agents

Key terms

reducing agent	antioxidant	free radical
oxidation, loss of electrons	oxidation, loss of hydrogen	oxidation, gain of oxygen
reduction, gain of electrons	reduction, gain of hydrogen	reduction, loss of oxygen
antioxidant vitamins	vitamin E	vitamin C
vitamin A	beta carotene	

8.4 Batteries

Batteries and electrochemical cells are defined. The components of a battery are described. Primary batteries, nonrechargeable ones, are defined and illustrated. Reusable, secondary, batteries are described. Reactions and applications are given for lead-storage batteries. Fuel cells are defined and applications for fuel cells are given. The hydrogen-oxygen fuel cell is diagrammed and described. Applications of fuel cells are discussed in electric automobiles, the Gemini, Apollo, and the space shuttle NASA programs.

Objectives
After studying this section a student should be able to:
- define electrochemical cell, anode, and cathode
- sketch and label a diagram for an electrochemical cell (see text Figure 8.5)
- describe the functions of anode, cathode, salt bridge
- define primary battery and secondary battery and give an example of each
- classify the following as either primary or secondary: lead-storage, Leclanché dry cell,
- describe the hazards posed by lead-storage batteries
- give the definition for a fuel cell and tell how they differ from batteries
- write the reaction that occurs in the hydrogen-oxygen fuel cell
- label a diagram of the hydrogen-oxygen fuel cell

Key terms

electrochemical cell	battery	two electrodes
anode	cathode	salt bridge
electric current	throw-away battery	primary battery
dry cell	renewable	secondary battery
lead-acid battery	recharging	short circuit
fuel cell	alkaline fuel cells	hydrogen-oxygen fuel cell
Gemini	Apollo	space shuttle
fuel cell stack		

8.5 Electrolysis: Chemical Reactions Caused by Electron Flow

This section defines electrolysis and explains how electrical energy is used to produce chemical change. Electroplating of copper and other metals is described. The electrolysis of water to produce hydrogen is summarized. Important chemicals produced by electrolysis are listed.

Objectives
After studying this section a student should be able to:
- give a definition for electrolysis
- describe what happens during electrolysis
- explain why aluminum, though abundant as a mineral, was difficult to purify

give an example of an electrolysis reaction and tell how it differs from a battery reaction
label a diagram of a copper electroplating cell and describe the electroplating process
name an important chemical substance prepared using electrolysis

Key terms

electrolysis reaction	electron flow	evolved
deposited	electroplating	plating

8.6   Corrosion--Unwanted Oxidation-Reduction

This section defines corrosion as the unwanted oxidation of metals. It describes in detail the most economically significant corrosion reaction, rusting. It gives the reactions for corrosion of iron and tells what conditions favor rust formation. The reactants needed for rusting: iron, oxygen and water are identified. Methods used to protect iron and prevent rusting are identified. A common way to protect iron with a zinc coating, galvanizing, is described. Some of its limitations are explained.

Objectives

After studying this section a student should be able to:
give a definition for corrosion
give a formula for rust and tell why it is a problem
tell what reactants are required for rust to form
explain how rusting can be prevented
describe what galvanizing is and how it prevents rusting
describe one way that galvanizing can fail

Key terms

corrosion	rust	salt bridge
electrochemical cell	protective coating	galvanizing

Additional Readings

Berendsohn, Roy & Rosario Copotosto. "Rust Prevention" <u>Popular Mechanics</u> Sept. 1993: 59.

Bergens, Steven H. et al. "A Redox Fuel Cell That Operates With Methane As Fuel At 120 °C" <u>Science</u> Sept. 2, 1994: 1418.

Consumer Reports Staff. "Best Bets in Auto Batteries" <u>Consumer Reports</u> Oct. 1995: 674.

Lave, Lester B. & Chris T. Hendrickson, Francis Clay McMichael. "Environmental Implications of Electric Cars" <u>Science</u> May 19, 1995: 993.

Lecard, Marc. "Recharge It" <u>Sierra</u> Nov.-Dec. 1993: 42.

Lipkin, Richard. "Firing Up Fuel Cells: Has a Space-age Technology Finally Come of Age for Civilians?" <u>Science News</u> Nov. 13, 1993: 314.

Moskal, Brian S. "It's the Battery, Stupid!" <u>Industry Week</u> Oct. 3, 1994: 22.

Vizard, Frank. "Strips of Power" <u>Popular Mechanics</u> July, 1994: 98.

© 1999 Harcourt Brace & Company. All rights reserved.

Answers to Odd Numbered Questions for Review and Thought

1.
   a. Oxidation is the loss of electrons such as in $Fe \longrightarrow Fe^{2+}(aq) + 2\ e^-$
   b. Reduction is the gain of electrons as in $F_2(g) + 2\ e^- \longrightarrow 2\ F^-$
   c. Oxidation is the gain of oxygen such as $2\ Ca(s) + O_2(g) \longrightarrow 2CaO(s)$
   d. Reduction is the gain of hydrogen as in $CH_2CH_2(g) + H_2(g) \longrightarrow CH_3CH_3(g)$
   e. Oxidizing agent is a substance that accepts electrons such as the Fe^{3+} in
      $Fe^{3+}(aq) + e^- \longrightarrow Fe^{2+}(aq)$
   f. Reducing agent is a substance that gives up electrons such as the Zn(s) in
      $Zn(s) \longrightarrow Zn^{2+}(aq) + 2\ e^-$

3.
   a. The carbon has more oxygens in CO_2 than in CO. Remember the O usually has an oxidation number of -2 in compounds. The oxidation number for carbon is +4 in CO_2 since oxygen has a -2 oxidation number. The two oxide ions O^{2-}, O^{2-} and C^{4+} combine for a total zero charge. Carbon has an oxidation number of only +2 in CO, since one O^{2-} is balanced by C^{2+}.
   b. The nitrogen has more oxygens in NO_2 than in NO. The oxidation number is +4 in NO_2 and only +2 in NO.
   c. The sulfur is more oxidized in SO_3 than in SO_2. The combined oxidation number for the three O's is -6 = -2 x 3 so the sulfur is +6 in SO_3. In the SO_2 the two O's have a combined oxidation number of -4 = -2 x 2 so the sulfur is + 4 in for SO_2.
   d. The chromium is more oxidized in CrO_3 than in CrO because it has more oxygens and has an oxidation number of +6 rather than +2.
   e. The calcium has an oxidation number of +2 in CaO and zero in the pure metal Ca. The Ca is combined with more oxygen in CaO than in the pure metal.
   f. The nitrogen in N_2 is more oxidized than the nitrogen in ammonia, NH_3, because it is combined with less hydrogen. The oxidation number is zero for nitrogen in N_2 but -3 in ammonia since each H has an oxidation number of +1.

5. There are two common oxides for hydrogen: water H_2O, and hydrogen peroxide, H_2O_2. The more common hydrogen oxide is water.

7. When hydrocarbons burn in a limited supply of oxygen carbon monoxide, CO is formed instead of carbon dioxide, CO_2. Less oxygen is needed to make a molecule of CO.

9.
   a. oxidation, Hydrogen is oxidized from a zero in pure hydrogen to 2+ in water.
   $$H_2(g) + O_2(g) \longrightarrow 2\ H_2O(l)$$
   Oxidation number is 0 ⎣_____⎦ Oxidation number is 1+
   Oxidation, loss of electrons
   Oxidation, gain of oxygen
   b. oxidation, Copper is oxidized from zero in pure metal to 2+ in CuO.
   $$2\ Cu(s) + O_2(g) \longrightarrow 2\ CuO(s)$$
   Oxidation number is 0 ⎣_____⎦ Oxidation number is 2+
   Oxidation, loss of electrons
   Oxidation, gain of oxygen

c. reduction, Tin, Sn, is reduced from a 2+ in SnO(s) to a zero in the pure metal.

$$2\ SnO(s) \longrightarrow 2\ Sn(s) + O_2(g)$$

Oxidation number is 2+ ⎿_____⏌ Oxidation number is 0
Reduction, gain of electrons
Reduction, loss of oxygen

d. reduction, Iron is reduced from a 3+ in Fe_2O_3 to a 0 in the pure metal.

$$Fe_2O_3(s) + 3\ C(s) \longrightarrow 2\ Fe(s) + CO(g)$$

Oxidation number is 3+ ⎿_____⏌ Oxidation number is 0
Reduction, gain of electrons
Reduction, loss of oxygen

e. reduction, Nitrogen is reduced from zero in the pure element to 3- in ammonia.

$$N_2(g) + 3\ H_2(g) \longrightarrow 2\ NH_3(g)$$

Oxidation number is 0 ⎿_____⏌ Oxidation number is 3-
Reduction, gain of electrons
Reduction, gain of hydrogen

f. oxidation, Sulfur is oxidized from a zero in the pure element to a 6+ in the SO_3 gas.

$$2\ S(s) + 3\ O_2(g) \longrightarrow 2\ SO_3(g)$$

Oxidation number is 0 ⎿_____⏌ Oxidation number is 6+
Oxidation, loss of electrons
Oxidation, gain of oxygen

11. Jupiter's atmosphere is reducing because methane, CH_4, and hydrogen, H_2, are abundant. This means there is a ready supply of hydrogen to act as a reducing agent.

13. The zinc is oxidized because Zn metal loses electrons when it forms the Zn^{2+} ion. The apparent charge or oxidation number increases from zero, 0, in the pure metal to +2 in the zinc cation. $Zn(s) \longrightarrow Zn^{2+}(aq) + 2\ e^-$

15. Corrosion of iron requires oxygen, O_2; water, H_2O; and iron, Fe. The rusting of automobiles is less of a problem in Arizona because the air is less humid there than in Chicago. There is less water in the air in Arizona.

17. By definition the anode is the electrode where oxidation occurs. To help keep this clear remember both anode and oxidation start with a vowel. a. True

19. Both a fuel cell and a battery use oxidation-reduction reactions to produce electrical energy. The reactants are added to a fuel cell and products are removed, while a battery is usually self-contained so that reactants and products are kept in the battery case.

21. See answer to question 19. The battery discharges to produce electricity and the contents of the battery are confined to the case of the battery. The reactants and products are still inside the battery. No material was added or removed and no mass was gained or lost.

© 1999 Harcourt Brace & Company. All rights reserved.

Speaking of Chemistry

Name _____

Oxidation-Reduction Reactions

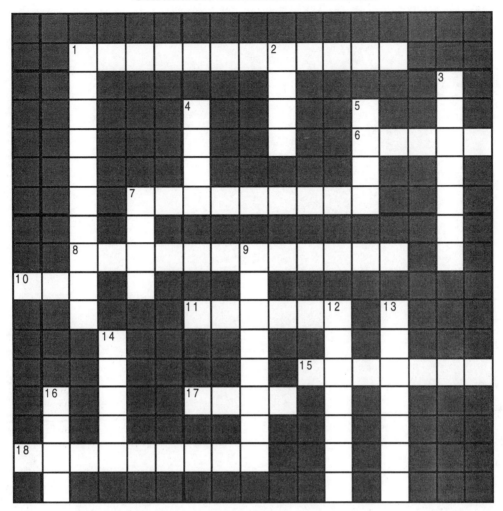

Across
1. Use of electrical energy to make a chemical reaction happen
6. Electrode where oxidation occurs
7. Gain of electrons
8. Oxidizing agents that kill microorganisms
10. Oxidation number on Na^{1+}
11. Chlorine acts to remove electrons from stain molecules, it is a ____
15. Electrolysis used to deposit metals
17. Oxidation number on atoms of pure elements
18. Unwanted redox reactions involving metals

Down
1. Anode and cathodes are
2. Oxidation is ____ of electrons, LOE
3. Atom or fragment with an unpaired electron
4. Metal in car batteries
5. Reduction is the ____ of electrons
7. Formed when iron corrodes
9. Oxidation is the loss of
12. Oxidizing agents in Group VIIA
13. Chemical source of electrical energy
14. ____ produced in H_2/O_2 fuel cell
16. Principle metal that undergoes corrosion

© 1999 Harcourt Brace & Company. All rights reserved.

Bridging the Gap
Silver Tarnish Removal by Oxidation - Reduction

Silver is a metal of great beauty and usefulness. Silver is used to make coins, jewelry, electrically conducting wire, and fine table utensils. Many of these articles are made from sterling silver, an alloy made from 7.5 parts copper and 92.5 parts silver. The copper increases the toughness of the metal. Silver coins need to be more robust than cake servers and punch bowls, so the coin silver used for quarters, dollars, etc. is an alloy of 10 parts copper and 90 parts silver.

Silver has a disadvantage. It tends to darken rapidly with an unattractive coat of tarnish on the metal surface. This brownish-black stain can be removed by rubbing with silver polish but this removes some of the metal as well as the tarnish. The tarnish is mainly silver sulfide, Ag_2S. The sulfur that reacts with the silver comes from a bad smelling toxic gas, hydrogen sulfide, H_2S. This is the gas that is responsible for the bad smell of a rotten egg. Hydrogen sulfide is emitted in some industrial processes and during the decomposition of plant and animal matter. Sulfur compounds are found in egg and mustard, so silver tarnish results from contact with these foods. Mayonnaise is made using whole eggs; it will cause silver to tarnish also.

An oxidation reduction reaction can be used to remove the sulfur from the tarnish and leave the silver intact. The silver in the silver sulfide, Ag_2S, can be reduced to silver metal by a more active metal. Aluminum is a more active metal than silver. This means that aluminum metal will react with silver ion. The aluminum metal, Al, will be oxidized to Al^{3+} while the silver ion, Ag^{1+}, is reduced to Ag metal.

This change can be described in terms of "oxidation numbers". An oxidation number is the apparent charge on a atom. The atoms in a sample of pure element always have oxidation number of zero. In an ionic compound, the oxidation number for each element is the same as its ion charge. For example, the sulfide ion, S^{2-}, in Ag_2S has a charge of 2- so its oxidation number is said to be -2. The sum of the oxidation numbers for a compound always equals zero.

Equipment and materials
A tarnish removal process is described below. To carry out this process you need the following equipment: aluminum foil, Pyrex™ glass container or Teflon™ lined pan large enough to hold the complete silver object, baking soda, tap water, heat resistant plastic spatula or a wooden utensil.
1. Make a baking soda solution by dissolving a heaping teaspoon of baking soda, $NaHCO_3$, in approximately one liter (roughly one quart) of water.
2. Use a microwave oven to heat the baking soda solution. The solution can be heated on a stove top burner in a Teflon™ lined pan or pot.
3. Place a piece of aluminum foil on the bottom of the bowl.
4. Be careful to avoid skin contact with the hot solution. Submerge the tarnished object completely in the solution.
5. Be sure the tarnished object is in contact with the aluminum foil. To speed up the process, use the wooden utensil to press the tarnished item against the aluminum. Notice that a reaction occurs almost immediately and gas bubbles form.
6. When the tarnish has been removed or there is no longer any significant change, remove the item from the solution and rinse it with clean tap water. Wipe the item dry with a soft cloth.

© 1999 Harcourt Brace & Company. All rights reserved.

Bridging the Gap   Name _____
Silver Tarnish Removal by Oxidation - Reduction

The combination of silver sulfide, aluminum metal and soda solution make an electrolytic cell. The aluminum is the negative electrode and it is oxidized to aluminum ions. The silver is the positive electrode where the silver ions are reduced to silver atoms. There is a passage of electricity and simultaneously some hydrogen gas is produced by the reduction of hydrogen in water to hydrogen gas.

1. **Nature of silver tarnish and Ag in sterling silver**
   What is the formula for tarnish on silver?

   What is the oxidation number for silver ion in tarnish?

   What is the oxidation number for silver in sterling silver?

   Is silver reduced when tarnish is converted to silver and the sulfide ion is carried into solution by H^{1+} ions from the baking soda? Explain.

2. **Nature of aluminum in foil in aluminum sulfide**
   What is the oxidation number for aluminum atoms in aluminum foil?

   What is the formula for the positive ion formed when aluminum metal is oxidized?
   (Hint: Remember aluminum is in Group IIIA.)

   What is the formula for the sulfide ion?
   (Hint: Sulfur is in GroupVIA and has six valence electrons.)

   Write the formula for aluminum sulfide.

3. **Equation for the oxidation-reduction reaction for tarnish removal**
   Write the reaction for the conversion of silver tarnish and aluminum metal to silver metal and aluminum sulfide.

4. **Prediction for tarnish removal**
   Copper tarnish is also a sulfide, CuS(s). Do you think copper tarnish can be removed using the same procedure used to remove tarnish from silver? Justify your answer. Think about the relative reactivity of copper and silver metals.

9 Energy and Hydrocarbons

9.1 Energy from Fuels

Combustion reactions are defined in this section. The origin of energy changes in combustion reactions is described in terms of the differences in bond energies for reactants and products. The heat of combustion calculation is illustrated using bond energies. Energy values per gram are given for common fuels.

Objectives

After studying this section a student should be able to:
- give definitions for the calorie and the joule
- give a definition for bond energy
- tell what determines whether a reaction takes in or gives off heat
- give a definition for heat of combustion
- calculate the energy change for a reaction using bond energy tables
- name the four classes of hydrocarbons

Key terms

fuel	calorie	joule
hydrocarbons	combustion calorimeter	heat of combustion
average bond energies	energy per gram	alkanes
alkenes	alkynes	aromatics

9.2 Alkanes--Backbone of Organic Chemistry

Alkanes (saturated hydrocarbons) are defined in this section. The importance of the tetrahedral shape around carbons in alkanes is reviewed. Detailed explanations of structural formulas, ball and stick models, and space filling models are given. Straight and branched chain alkanes are defined. The differences between straight and branched chain alkane isomers are illustrated.

Objectives

After studying this section a student should be able to:
- give a definition for saturated hydrocarbons and alkanes
- write the general formula for alkanes
- sketch the tetrahedral structure for methane and give the bond angles
- describe how boiling points for alkanes change with increased carbon chain length
- sketch a structural formula, ball-and-stick formula and space-filling model for methane and for ethane
- tell how straight-chain isomers differ from branched-chain isomers
- give a definition for isomers and give examples
- write the structural formulas for the isomers of C_4H_{10}
- give names for common alkanes
- tell what is meant by the term alkyl group
- give structures for the following alkyl groups:
    methyl, ethyl, propyl, isopropyl, butyl, t-butyl
- describe the naming system for alkyl groups and how they relate to alkane names

© 1999 Harcourt Brace & Company. All rights reserved.

Key terms

alkane	general formula C_nH_{2n+2}	carbon-carbon single bond
saturated hydrocarbons	tetrahedral	ball-and-stick model
space-filling model	straight-chain	branched-chain
structural formula	isomers	structural isomers
methyl group	alkyl groups	alkane names
alkyl group names	"-ane" suffix	"-yl" suffix

9.3 Alkenes and Alkynes: Reactive Cousins of Alkanes

Alkenes are defined and their economic importance is described in this section. The general formula, C_nH_{2n}, for alkenes is given. The reactivity of alkenes is described. Structural isomers and stereoisomers are defined. Examples of *cis* and *trans* isomers for alkenes are defined and illustrated. Alkynes are defined in this section. The general formula for alkynes, C_nH_{2n-2}, is given. The linear structural characteristic around the triple bond in alkynes is illustrated. The basic naming system for alkynes is introduced.

Objectives
After studying this section a student should be able to:
- give the definitions for alkenes and alkynes
- write the general formula for alkenes with one double bond
- draw the structural formulas for ethene and for propene
- tell what bond angles are characteristic for alkenes and sketch examples of alkenes
- describe *cis* and *trans* isomers and give examples
- write the general formula for alkynes with one triple bond
- draw the structural formula for ethyne
- tell what bond angle is characteristic of an alkyne

Key terms

alkene	general formula C_nH_{2n}	carbon-carbon double bond
alkene names	numbering the carbon chain	naming structural isomers
"-ene" suffix	propylene	structural isomers of alkenes
cis isomer	*trans* isomer	stereoisomerism
optical isomerism	free rotation	degree of flexibility of bonds
alkynes	general formula C_nH_{2n-2}	$-C \equiv C-$
"-yne" suffix	naming alkynes	carbon-carbon triple bond

9.4 The Cyclic Hydrocarbons

Cycloalkanes, cyclopropane, cyclobutane, cyclopentane and cyclohexane described and illustrated in this section. Aromatic compounds are defined and the "smearing" of electrons around the benzene ring is explained. Substituted derivatives of benzene and the naming system for disubstituted isomers are shown. Polycyclic aromatics like naphthalene and anthracene are described. Examples of steroid structures that incorporate cyclic groups are illustrated.

Objectives
After studying this section a student should be able to:
- explain how cyclic alkanes differ from straight-chain hydrocarbons
- draw structures for cyclopropane, cyclobutane, cyclopentane
- give the definition for aromatic compounds and draw the structural formula for benzene
- draw the structures for meta-, para- and ortho-dichlorobenzene, $C_6H_4Cl_2$

Key terms
aromatics
benzene, C_6H_6
ortho-
naphthalene
natural product chemists
cycloalkanes
planar structure
meta-
anthracene
benzene
derivatives of benzene
para-
steroids

9.5 Alcohols: Oxygen Comes on Board

Alcohol, ROH, and ether, ROR', functional groups are introduced. Oxygen-containing functional groups with single bonds like alcohols and ethers are described. Wood and grain alcohol are discussed in detail. The reactions used in the production of alcohols are summarized. Gasohol and oxygenated gasolines are described. Ethers, and their applications are described. The use of MTBE, methyl-tertiary-butyl ether as a fuel additive is explained. MTBE

Objectives
After studying this section a student should be able to:
  give the definition for a functional group
  give definitions for alcohols and ethers
  write the formulas for ethanol and methanol
  describe how methanol is produced
  describe how ethanol can be produced
  give uses for ethers
  tell what MTBE is and describe its major use
  give the definition for gasohol and two reasons for its use

Key terms
functional groups
methanol and CH_3OH
ethers, R-O-R'
diethyl ether, $C_2H_5OC_2H_5$
alcohols, R-OH
ethanol and C_2H_5OH
anesthetic
methyl propyl ether
fermentation
gasohol
octane enhancer, MTBE
$CH_3OCH_2CH_2CH_3$

9.6 Petroleum

Petroleum and its sources are described in this section. The growth in petroleum consumption and the estimates of global reserves are described. Fractional distillation and petroleum refining are explained and illustrated. Petroleum fractions and their uses are tabulated. Octane ratings for gasoline are explained. The standard, 2,2,4-trimethylpentane, for octane rating and its relation to an octane rating of 100 are discussed. Octane enhancers and gasoline additives are tabulated. Catalytic re-forming of straight chain hydrocarbons is described. Reformulated and oxygenated gasolines are discussed. The catalytic cracking process is described. Cities across the USA that are required to use oxygenated gasoline during winter are identified.

© 1999 Harcourt Brace & Company. All rights reserved.

Objectives
After studying this section a student should be able to:
  tell how long known oil reserves are projected to last at current consumption levels
  describe the process for refining petroleum by fractional distillation (use a sketch)
  give a definition for a petroleum fraction
  tell what is meant by straight-run gasoline
  give the definition of octane rating
  name the compound used to define the 100 octane rating
  tell how the octane rating for a fuel is determined
  tell how the octane rating for a fuel is related to regular, regular plus and premium classes
  explain why knocking and pinging occur in an automobile engine
  explain why catalytic re-forming of hydrocarbons is done
  describe the catalytic cracking process
  give definitions for octane enhancers, reformulated gasoline, and oxygenated gasoline
  tell why alkenes and aromatic compounds are added to straight-run gasoline

Key terms

crude petroleum	fractional distillation	distillation tower
petroleum fractions	volatile	condense
lower-boiling fractions	higher-boiling fractions	"straight-run"
residual oil	octane rating	isooctane
"knocking"	"pinging"	reformulated gasoline
premium gasoline	regular plus gasoline	catalytic re-forming process
octane enhancers	tetraethyl lead	straight-chain hydrocarbons
MTBE	catalytic cracking process	branched-chain hydrocarbons
oxygenated gasoline		

9.7 Natural Gas

The composition of natural gas is described as a mix of C_1 to C_4 hydrocarbons. The main component of natural gas is methane, CH_4. Current use and projected U.S. consumption of natural gas are discussed. Efforts to use natural gas as a vehicle fuel are described.

Objectives
After studying this section a student should be able to:
  describe the composition of natural gas
  describe the uses for natural gas
  explain the benefits and disadvantages associated with using natural gas as a vehicle fuel

Key terms

natural gas	C_1 to C_4 alkanes	methane
liquefied gas	butane	propane
benefits of vehicle fuel	organic chemical industry	energy source

9.8 Coal

The composition of coal is described as a mixture of hydrocarbons with more fused rings than petroleum. Electricity production is identified as the main use for coal. The current levels of coal production and coal reserves are discussed. Coal gasification to produce synthesis gas and methane are explained using reactions. Coal liquefaction using pressurized hydrogen and catalysts to produce a fluid crude oil-like product is described.

Objectives
After studying this section a student should be able to:
- describe the composition of coal
- tell why the use of coal as a heating fuel decreased
- tell how synthesis gas is produced
- describe the coal gasification process to make methane
- explain how coal gasification differs from coal liquefaction

Key terms
- coal
- deep coal mining
- synthesis gas
- coal reserves
- strip mining
- coal liquefaction
- fused rings of carbon atoms
- coal gasification
- hydrogenating coal

9.9 Methanol as a Fuel
This section describes programs to replace gasoline with methanol and other fuels. Flexible-fueled vehicles, FFVs, are defined. Definitions are given for alternate fuels and mixtures. M100 is pure methanol and M85 is a blend of 85% methanol with 15% gasoline. Advantages and disadvantages of methanol and alternate fueled vehicles are discussed. The New Zealand program that produces gasoline from methanol is described.

Objectives
After studying this section a student should be able to:
- give reasons for using methanol as a replacement fuel
- explain why FFVs are being developed
- tell what the abbreviation code names for fuels like M100, E100, and M85 mean
- give two advantages that methanol offers as a fuel
- describe two problems that methanol presents as a fuel

Key terms
- methanol
- ethanol
- FFV, flexible-fueled vehicles
- M100
- E100
- exhaust emissions
- M85
- E85

Additional Readings

Anderson, E.V. "Ethanol's Role in Reformulated Gasoline Stirs Controversy" <u>Chemical & Engineering News</u> Nov. 2, 1992: 7.

Anderson, E.V. "Health Studies Indicate MTBE Is Safe Gasoline Additive" <u>Chemical & Engineering News</u> Sept. 20, 1993: 9.

Bowler, Sue. "Where the Power Lies" <u>New Scientist</u> Jan. 23, 1993: 32.

Emsley, John. "Energy and Fuels" <u>New Scientist</u> Jan. 15, 1994: 51.

Kerr, Richard A. "U. S. Oil and Gas Fields Double in Size" <u>Science</u> Feb. 24, 1995: 1090.

Lewis, Bernard. "High Temperatures: Flame" <u>Scientific American</u> Sept. 1954: 84.

Peaff, George "Court Ruling Spurs Continued Debate Over Gasoline Oxygenates" <u>Chemical & Engineering News</u> Sept. 26, 1994: 8.

© 1999 Harcourt Brace & Company. All rights reserved.

Rogge, Wolfgang F. et al. "Natural Gas Home Appliances (ES &T Research: Sources of Fine Organic Aerosol, part 5)" <u>Environmental Science & Technology</u> Dec. 1993: 2736.

Spencer, Brian. "Hydrocarbon Fuels, No Thanks" <u>New Scientist</u> Mar. 18, 1995: 45.

Answers for Odd Numbered Questions for Review and Thought

1. Fossil fuels are coal, crude oil, natural gas or heavy oil formed over millions of years by the action of heat and pressure in the earth's crust on buried decayed animal and plant matter.

3. An hydrocarbon is a compound such as an alkane, 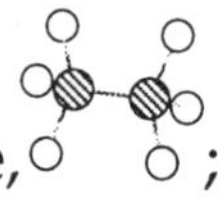 ; alkene, 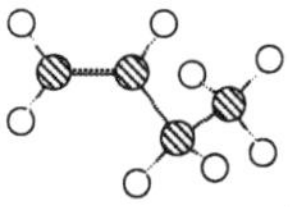 ; or alkyne,  ; containing only carbon and hydrogen.

5. The heat of combustion is the energy released when a compound reacts completely with oxygen.

7. a. Isomers are molecules with the same general formula but different molecular structures, like C_5H_{12} with isomers $CH_3CH_2CH_2CH_2CH_3$ and $CH_3CHCH_2CH_3$ with CH_3 branch
    b. Straight-chain hydrocarbons are compounds with a linear chain of CH_2-units and terminal CH_3 units like normal butane, 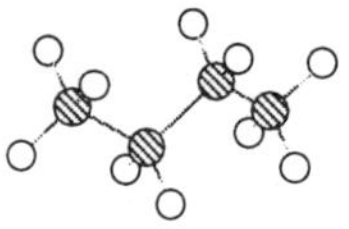
    c. Branched-chain hydrocarbons have a backbone like a straight-chain hydrocarbon, but they have some H's along the chain replaces by alkyl groups, C_nH_{2n+1}.

9. The first four members of the alkane series all have single bonds between the carbon atoms and the maximum number of hydrogen atoms.

    n=1, methane, CH_4,

    n=2 ethane, C_2H_6

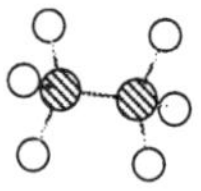

    n=3 propane, C_3H_8

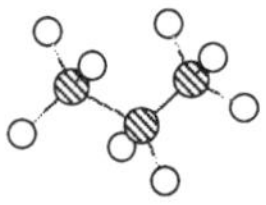

    n=4 butane, C_4H_{10}

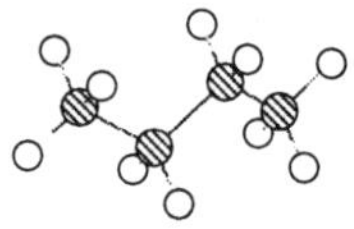

11. a. 2-Methylpentane, $CH_3CHCH_2CH_2CH_3$ with CH_3 branch
    b. 4,4-Dimethyl-5-ethyloctane, $CH_3CH_2CH_2CCHCH_2CH_2CH_3$ with CH_3, H_3C, CH_2CH_3 branches
    c. 2-Methyl-2-hexene, $CH_3C=CHCH_2CH_2CH_3$ with CH_3 branch

138
© 1999 Harcourt Brace & Company. All rights reserved.

13. There are three structural isomers for C₅H₁₂. There is no simple way to determine the number of such isomers. The only way to do this is to follow a systematic set of steps. The first thing to do is to draw the structure for the straight chain isomer. The next is to shorten the chain by one carbon and attach this carbon to the number two carbon on the chain. The "moving" or "attaching" carbon atom is then positioned on the next carbon and so forth until duplicate structures are produced. After all these possibilities are exhausted the carbon chain is shortened by another carbon atom and the possible combinations are again tried. There are only the following three different pentane isomers.

pentane	2-methylbutane	2,2-dimethylpropane
CH₃CH₂CH₂CH₂CH₃	CH₃CHCH₂CH₃ with CH₃ branch	CH₃CCH₃ with two CH₃ branches

15.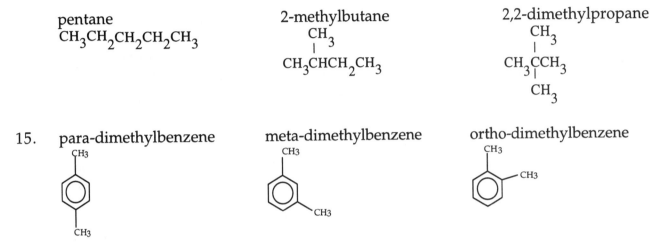
   para-dimethylbenzene, meta-dimethylbenzene, ortho-dimethylbenzene

17. There are *cis-* and *trans-* isomers for 2-butene because there are carbon chains on both double-bonded carbon atoms. These can be positioned either on the same side of the double bond, cis, or on opposite sides, trans.

    trans isomer   cis isomer

    The 1-butene has only hydrogens on the #1 carbon so there is no group that can be cis or trans to the groups on the #2 carbon.

19. Hot petroleum at about 400 °C is introduced into a fraction distillation tower with collection trays at different temperatures. Each component will condense at a level with the range of boiling temperatures and condensation temperatures that match the temperature of a specific tray. The compounds that are most volatile with the lowest boiling points are collected near the top. The compounds like tars that are least volatile with the highest boiling point are collected at the bottom.

21. Aromatic hydrocarbons, branched-chain hydrocarbons, octane enhancers like toluene, 2-methyl-2-propanol, methanol, ethanol, or methyl-tertiary-butyl ether

23. Synthesis gas, a mixture of CO and H_2, is produced by passing super-heated steam over pulverized coal.
    $$C + H_2O + 31 \text{ kcal} \longrightarrow CO + H_2$$
    Coal gasification using a catalyst, crushed coal and synthesis gas will produce methane.
    $$2 C + 2 H_2O + 2 \text{ kcal} \longrightarrow CH_4 + CO_2$$

25. Methanol-to-gasoline is possible through a multi-step catalyzed process.
    $$2 CH_3OH \xrightarrow{\text{ZSM-5 catalyst}} (CH_3)_2O + H_2O$$
    $$2 (CH_3)_2O \xrightarrow{\text{ZSM-5 catalyst}} 2 C_2H_4 + 2 H_2O$$
    $$2 C_2H_4 \xrightarrow{\text{ZSM-5 catalyst}} \text{gasoline, a mixture of } C_5\text{-}C_{12} \text{ hydrocarbons}$$

27. Legal requirements for enhanced oxygenated gasoline will increase demand for methanol. The social and environmental need for cleaner burning gasoline will likely lead to more methanol use. Gasoline refined from petroleum will probably increase in price as oil reserves decline. This will increase the cost effectiveness of methanol as a fuel.

29. Oxygenated gasolines cause less pollution because they burn more completely. The presence of more oxygen in the combustion mixture makes the combination of oxygen with carbon more complete. This is expected to reduce hydrocarbon exhaust emissions, carbon monoxide levels and ozone emissions.

31. Treating pulverized coal with superheated steam forms either carbon monoxide and hydrogen or, using a catalyst, carbon dioxide and methane.

33. The M identifies a fuel blend of methanol and gasoline. The E identifies a fuel blend of ethanol and gasoline. The number following the symbol tells the percentage of these alcohols in the mixture. The remaining component in the blend is gasoline.
    M100 means 100% methanol.
    E100 means 100% ethanol.
    M85 means 85% methanol, 15% gasoline.
    E85 means 85% ethanol, 15% gasoline.
    FFV means flexible-fueled vehicle which can operate on gasoline or other fuels.

35. a. methane, alkane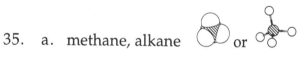

    b. benzene, aromatic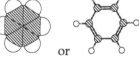

    c. 1-butene, alkene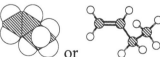

    d. acetylene, alkyne

Speaking of Chemistry

Name _____

Energy and Hydrocarbons

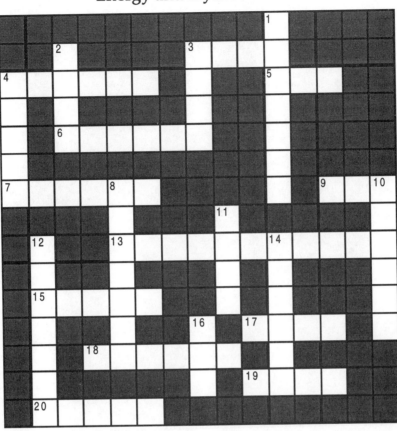

Across
3 number of carbons in butane
4 measures gasoline knocking behavior
5 common word for petroleum
6 C_2H_6 or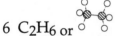
7 methanol adds _____ to gasoline
9 alkene with substituents on same side
13 contain only H, C
15 alkene isomer like
17 Aromatic compound with groups in the 1 and 4 positions,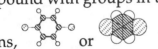
18 have general formula C_nH_{2n}
19 prefix used to indicate the position of substituents in an aromatic molecule like.
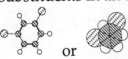
20 Oxygen containing compound with the formula $CH_3CH_2OCH_2CH_3$.

Down
1 Hydrocarbon with formula $CH_3CH_2CH_3$. or
2 Octane enhancer, methyl-tertiary-butyl ether.
3 Number of carbons in pentane
4 The prefix used to indicate the position of substituents in an aromatic molecule like.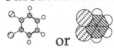
8 CH_3CH_2OH
10 kind of bonds in saturated hydrocarbons
11 Solid mixture of hydrocarbons and small amounts of sulfur.
12 main component in natural gas, CH_4
14 hydrocarbons with formula C_nH_{2n+2}
16 number of carbon atoms in methane.

© 1999 Harcourt Brace & Company. All rights reserved.

Bridging the Gap
Vehicle Emissions Tests: Reading the Results

A sample emissions test report is illustrated below. This type of report is required in many areas of the country before a vehicle can be licensed. This exercise is intended to familiarize you with the kinds of information that appears on one of these test forms. The emissions test looks at hydrocarbons, HC; carbon monoxide, CO; and the sum of carbon monoxide and carbon dioxide, $CO + CO_2$. The form shows the standards for the specific vehicle. The actual results are printed below the standard allowed emissions. The allowed HC emission value is 220 parts per million. The actual value is only 3 PPM so the vehicle passed and it is burning gas relatively efficiently. The carbon monoxide standard is 1.2%. The actual value is 00.00%. This means the vehicle is not emitting any toxic CO. The minimum level of carbon monoxide and carbon dioxide is 6.00%. The actual test result is 15.10% which is consistent with efficient burning of gasoline and a properly working catalytic converter. This vehicle passed in all three categories. This is indicated in by the big PASS opposite the final results arrow as well as the pass symbols under the separate results. A similar test report sheet appears on the following page. Your assignment is to determine whether or not the second vehicle passed. You are supposed to tell how well the vehicle met the standards, or, if it failed, tell what standards were not met. You also are supposed to tell what changes in emissions might be needed.

VEHICLE EMISSIONS TEST REPORT   6726078
This Form is Your Receipt for the Test Fee   6726078

This test was performed in accordance with federal regulations on emission tests (40 CFR 85 Subpart W)

Inspection information

License	Vehicle number	Year	Make	FINAL RESULTS	PASS
ACS 139	4S2CY45V783200172	93	HONDA		

VEHICLE INFORMATION

G.V.W.R.	ODOMETER x 1000	FUEL TYPE	TEST NO
	34	G	1

INSPECTION STATION INFORMATION

INSPECTOR.	STATION	MO. DAY YR	HR. MIN.
138	05	2 12 94	13 24

GASOLINE EMISSION TEST RESULTS

TEST: TWO SPEED	HC (PPM)	CO (%)	$CO + CO_2$ (%)	OPACITY(%)
Emissions Limits	0220	01.20	≥6.00	
Emissions at 0749 RPM	0003	00.00	15.10	
Emissions at 2256 RPM	0007	00.00	15.24	
	P	P	P	
RESULTS: PASS: (P) FAIL: (F)				

© 1999 Harcourt Brace & Company. All rights reserved.

Bridging the Gap
Concept Report Sheet   I   Name _____

Use the sample test report below to answer the following questions.

Did this vehicle pass the emissions test?

How well did the hydrocarbon emissions meet the emissions limits?

Is the engine doing a good job of completely burning the gasoline? Explain.

How well did the vehicle meet the CO emission limits? Does the catalytic converter appear to be working well?

VEHICLE EMISSIONS TEST REPORT   6726092
This Form is Your Receipt for the Test Fee   6726092

This test was performed in accordance with federal regulations on emission tests (40 CFR 85 Subpart W)

Inspection information

License	Vehicle number	Year	Make	FINAL RESULTS	
ACS 719	7S2CY45V783200172	89	FORD		FAIL

VEHICLE INFORMATION

G.V.W.R.	ODOMETER x 1000	FUEL TYPE	TEST NO
	64	G	1

INSPECTION STATION INFORMATION

INSPECTOR.	STATION	MO. DAY YR	HR. MIN.
121	06	4 19 96	11 44

GASOLINE EMISSION TEST RESULTS

TEST: TWO SPEED	HC (PPM)	CO (%)	CO + CO_2 (%)	OPACITY(%)
Emissions Limits	0220	01.20	≥6.00	
Emissions at 0749 RPM	0340	04.00	3.20	
Emissions at 2256 RPM	0422	08.00	3.80	
	F	F	F	
RESULTS: PASS: (P) FAIL: (F)				

10 Organic Chemicals and Polymers

10.1 Organic Chemicals
This section describes the sources of organic chemicals. It tabulates the classes of compounds produced by distillation of coal tar. The concept of functional groups is introduced.

Objectives
After studying this section a student should be able to:
- tell how coal gas is produced
- identify the process used to make ethylene, propylene, etc. from petroleum
- tell what the term functional group means

Key terms
- organic chemicals
- aromatic compounds
- hydrocarbons
- functional groups

10.2 Alcohols and Their Oxidation Products
The characteristic functional group, -OH, for alcohols is identified. The general formula for alcohols ROH and many specific examples are given. Sources and methods to produce ethyl alcohol are discussed. Primary, secondary, and tertiary alcohols are defined and some important alcohols are described. Oxidation reactions of alcohols to form aldehydes, ketones, and carboxylic acids are illustrated. Hydrogen bonding in alcohols is detailed. Alcohol, ethanol, toxicity, alcohol proof, and denatured alcohol are discussed. Ethylene glycol and glycerol are described.

Objectives
After studying this section a student should be able to:
- tell how primary, secondary and tertiary alcohols differ
- describe oxidation reactions of alcohols to form aldehydes, ketones & carboxylic acids
- tell how hydrogen bonding in alcohols arises
- tell how hydrogen bonding differs in alcohols with one, two, and three -OH groups
- explain how proof and percent alcohol by volume are related to each other
- convert percent alcohol by volume to proof & proof to percent alcohol
- tell why and how alcohol is denatured
- describe the difference between ethanol and ethylene glycol
- classify a compounds as an aldehyde, ketone, alcohol, carboxylic acid from the formula

Key terms

primary alcohol	secondary alcohol	tertiary alcohol
"R- group"	R- and R'-	structural isomers
oxidation of alcohol	ethanol	hydrogen bonding
solubility of alcohols	formaldehyde, methanal	fermentation
"proof"	ethanol metabolized	denatured alcohol
detoxification	alcohol, "-ol"	acetic acid
ethylene glycol	permanent antifreeze	glycerol ("glycerin")
carcinogen	aldehyde, "-al"	carboxylic acid, "-oic"

acetaldehyde,

© 1999 Harcourt Brace & Company. All rights reserved.

10.3 Carboxylic Acids and Esters

This section describes the classes of compounds containing a carbon atom doubly bonded to an oxygen atom: aldehydes, ketones, carboxylic acids and esters. Common aldehydes and ketones are described. Simple carboxylic acids and naturally-occurring carboxylic acids are discussed. The preparation and properties of esters are also described.

Objectives
After studying this section a student should be able to:
- classify a compound as either a carboxylic acid or an ester
- give a brief description of the properties of esters and carboxylic acids
- write the general formula for each of the following and circle the functional group: carboxylic acids, esters
- tell what the natural sources are for citric acid, lactic acid, stearic acid, oleic acid
- tell what alcohol and acid are needed to prepare an ester given the name of the ester
- given the structural formula for an ester, tell what alcohol and acid were used to make it
- given an alcohol and an acid, draw the structural formula of their ester

Key terms
carbonyl group	-COOH functional group	RCOOH
formic acid, HCOOH	dicarboxylic acid	ester, "-ate", RCOOR'
acetic acid, CH_3COOH	flavors	food additives

10.4 Organic Chemicals from Coal

This section describes how organic chemicals can be made from coal and steam. Specific reactions are illustrated for the production of synthesis gas, methanol, methyl acetate, and acetic anhydride.

Objectives
After studying this section a student should be able to:
- write the reaction for preparation of synthesis gas
- identify the structural formulas for methanol, methyl acetate, and acetic anhydride
- tell what products result from the reaction of water and acetic anhydride

Key terms
chemicals from coal	gasification of coal	synthesis gas
methanol	acetic acid	acetic anhydride
methyl acetate		

10.5 Synthetic Organic Polymers

Definitions are given for plastics, monomers, polymers, addition polymers, and condensation polymers. Addition polymers such as polyethylene, and polymers derived from polyethylene-like compounds are described. Conditions for cross-linked polyethylene, CLPE, and branched polyethylene, LDPE, are described. Poly(vinyl chloride), PVC, synthesis is outlined. Natural rubber (polyisoprene) and synthetic rubber (polybutadiene) structures and properties are illustrated and discussed. Vulcanization of natural rubber is explained. Stereoregulation in synthetic rubber is described. The structure of neoprene rubber is illustrated. Copolymers are defined and styrene-butadiene rubber, SBR, is illustrated. Condensation polymers such as polyesters(PET) and (Dacron™), and polyamides (nylon), are described.

Objectives
After studying this section a student should be able to:
- give definitions for monomer and polymer
- tell how addition polymers differ from condensation polymers

give definitions for free radical and initiation
tell how the polyethylene chain is formed from ethylene
give definitions for HDPE, LDPE, and CLPE
write the formula for styrene and for polystyrene
identify the monomer that produces a specific polymer
identify monomers such as ethylene, vinyl chloride, styrene, etc. if given the structures,
tell how and why vulcanized rubber differs from natural rubber
describe a condensation reaction
identify the representative formula for nylon-66

Key terms

polymer	monomers	addition polymers
initiation	free radical	ethylene, $CH_2 = CH_2$
polyethylene	HDPE	LDPE
CLPE	styrene	polystyrene
polypropylene	poly(vinyl chloride), PVC	natural rubber
vulcanized rubber	isoprene	latex
poly-*cis*-isoprene	elastomers	polybutadiene
stereoregulation	gutta-percha	*trans* polymer
copolymers	synthetic rubber	neoprene
condensation reaction	polyesters	styrene-butadiene, SBR
condensation polymers	polyamides	PET
amide linkage	nylon-66	

10.6 New Polymer Materials

This section describes reinforced plastics. The properties of reinforced plastics are compared with the properties of metals. Glass fiber and carbon fiber composites are discussed. Applications of composite materials are described.

Objectives

After studying this section a student should be able to:
give a definition for composites and reinforced plastics
describe how the strength and density of composites compare with those of aluminum and steel
give a definition for specific gravity
name the fiber most commonly used in composites
describe how graphite/polymer composites are constructed
name applications for glass fiber and graphite/polymer composites

Key terms

reinforced plastics	composites	polymer matrix
low specific gravity	high resistance	chemical resistance
glass fibers	polyesters	graphite fibers
graphite-epoxy composites		

10.7 Recycling Plastics

This section describes the present level of plastics recycling and tells why metals such as lead are extensively recycled. The four phases for successful waste recycling are identified.
collection   sorting   reclamation   end use
The code system for plastic identification is tabulated. The most commonly recycled plastics, PET and HDPE, are described.

© 1999 Harcourt Brace & Company. All rights reserved.

Objectives
After studying this section a student should be able to:
describe the main reason why plastics recycling lags behind recycling of lead and iron
identify the four phases needed for successful recycling of any waste
describe how plastic labeling aids consumers in identification of plastics for recycling
identify the two most recycled plastics and name the recycled applications

Key terms

recycling plastics	number-one waste	collection
sorting	reclamation	end-use
PET	HDPE	most commonly recycled plastic
resin	poly(ethylene terephthalate)	

Additional Readings

Fisher, Harry L. "Rubber" <u>Scientific American</u> Nov. 1956: 74. I

Natta, Giulio. "How Giant Molecules Are Made" <u>Scientific American</u> Sept. 1957: 98.

Oster, Gerald. "Polyethylene" <u>Scientific American</u> Sept. 1957: 139.

Wingo, Walter. "Government Plans to Bolster 'Green' Design. (National Environmental Technology Strategy Promotes Environmentally-Friendly Technologies)" <u>Design News</u> June 26, 1995: 39.

Answers to Odd Numbered Questions for Review and Thought

1. Carbon atoms can bond to other carbon atoms in almost unlimited numbers and in a variety of ways. Introducing other elements allows different molecules due to different atom sequences (functional groups and isomers).

3. a. carboxylic acid,

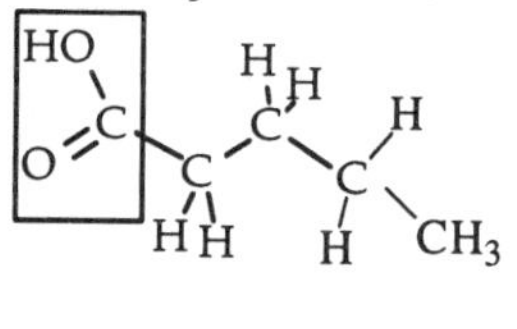

    b. alkene,

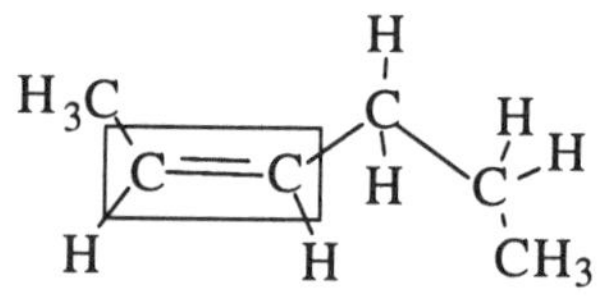

    c. symmetric ether,

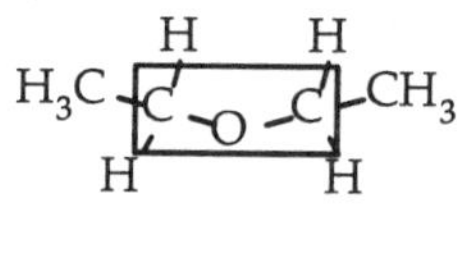

    d. secondary alcohol,

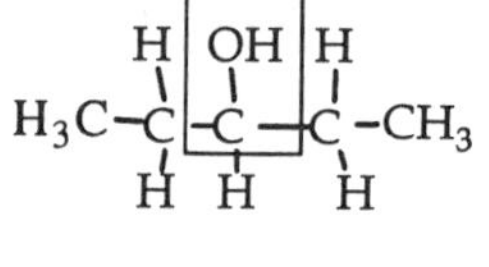

    e. tertiary alcohol,

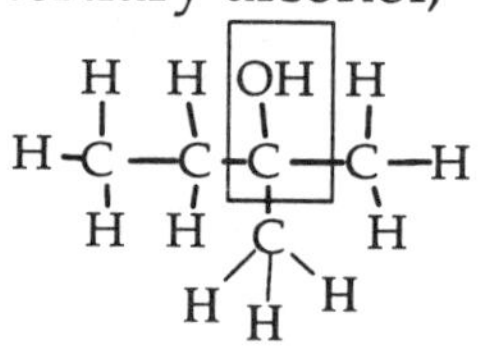

    f. aldehyde,

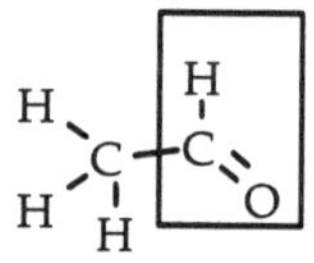

5. a. carboxylic acid, pentanoic acid

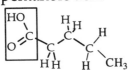

b. symmetric ether, dimethyl ether

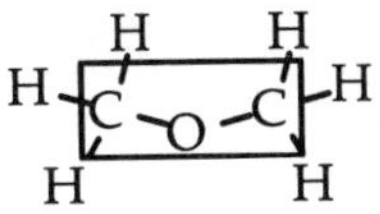

c. secondary alcohol, 2-butanol

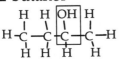

d. a ketone, acetone

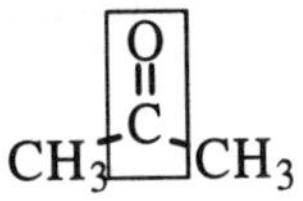

e. aldehyde, propanal

f. an ester, ethyl acetate

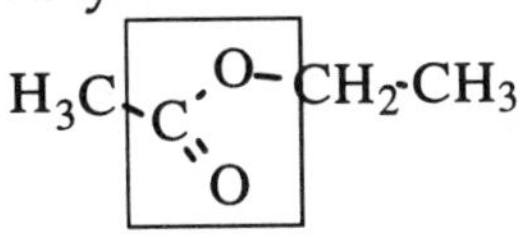

H₃C–C(=O)–O–CH₂CH₃

7. Proof = 2 × percent alcohol by volume   This means an 80 proof liquor is really 40% alcohol by volume and 60% water. Similarly 100 percent alcohol is 200 proof.

90 proof = $\frac{90}{2}$% = 45%

9. ethanol

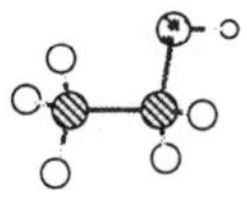

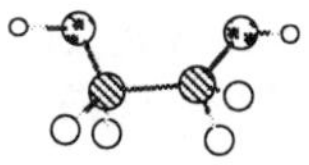

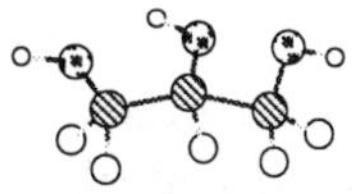

CH₃CH₂OH
one -OH group
use: solvent

ethylene glycol

CH₂OHCH₂OH
two -OH groups
use: automobile antifreeze

glycerol

CH₂OHCHOHCH₂OH
three -OH groups
use: cosmetics

11. a. aldehyde group

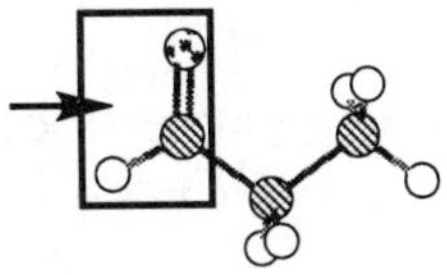

b. alcohol group

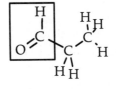

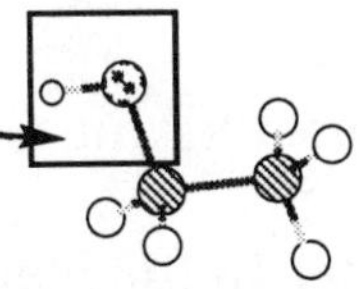

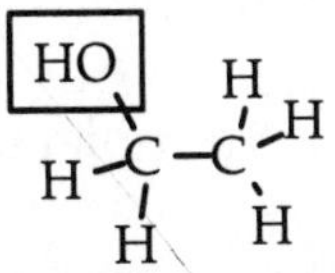

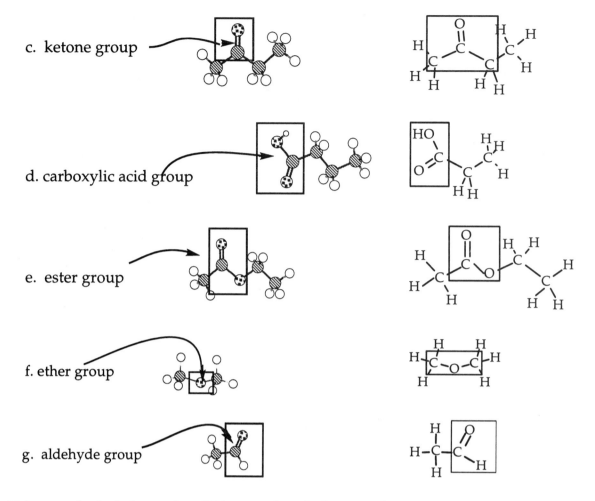

c. ketone group

d. carboxylic acid group

e. ester group

f. ether group

g. aldehyde group

13. Primary alcohols have the -OH group bonded to a carbon atom attached to one other carbon. Secondary alcohols have the -OH group bonded to a carbon atom attached to two other carbon atoms. Tertiary alcohols have the -OH group attached to a carbon atom that is bonded to three other carbon atoms.

15. The liver detoxifies ethanol by oxidation. Ethanol is oxidized to acetaldehyde which is oxidized to acetic acid. This is further oxidized to CO_2 and H_2O.

17. A railroad train has a series of repeating, identical, linked units, the railroad cars. Polystyrene is a long chain of repeating, identical, linked units, the styrene molecules.

19. A macromolecule is a molecule with very high molecular mass.

21. Monomers need to contain a multiple bond in order to form addition polymers. The double bond is converted to a single bond. A triple bond is converted to a double bond. The addition polymer adds units at the expense of one of the bonds in the multiple bond.

23. Natural rubber is poly-*cis*-isoprene. The individual monomer is isoprene.

isoprene

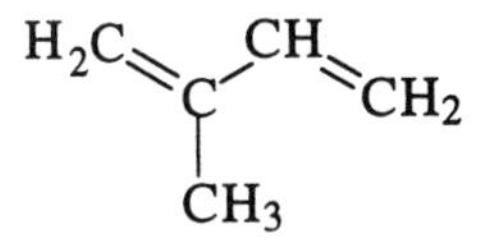

poly-*cis*-isoprene

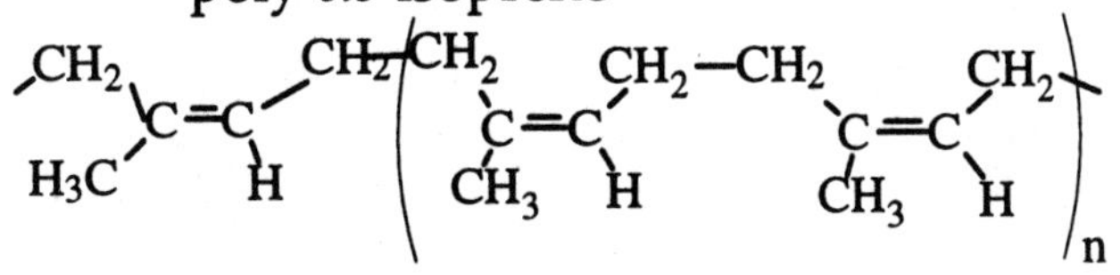

25. Sulfur is reacted with rubber used to align rubber polymers and bond the polymer chains to one another. The sulfur cross links toughen the rubber.

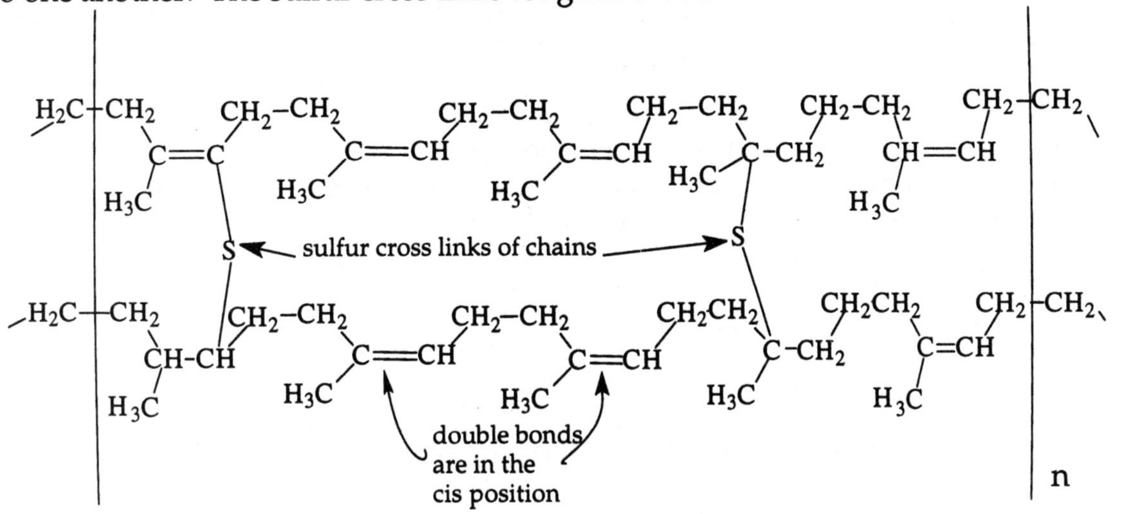

27. Cross linking makes the polymer more rigid.

29. a. Yes, styrene can undergo an addition reaction because it can add to double bond in the ethene group. The addition converts the double bond into a single bond. An atom or group adds to both ends of the double bond. This process is illustrated here.

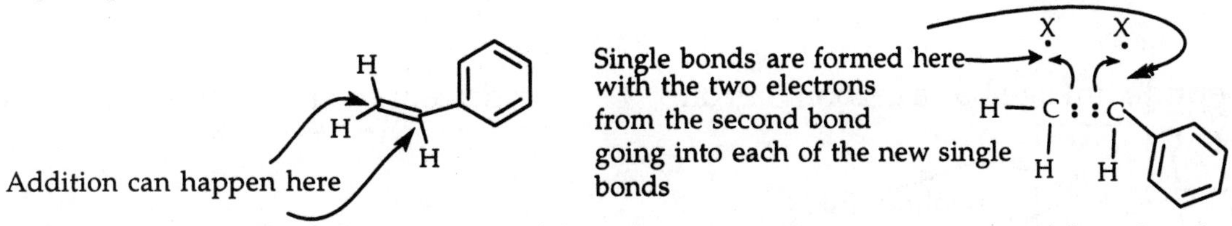

b. Yes, propene can undergo an addition reaction because it can add to the double bond.

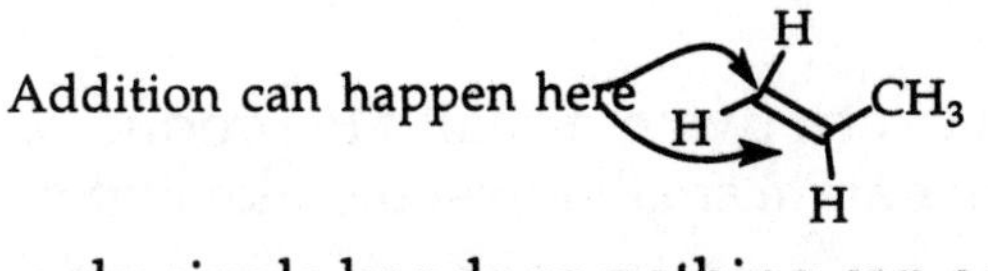

c. No, there are only single bonds so nothing can add to any multiple bonds in ethane.

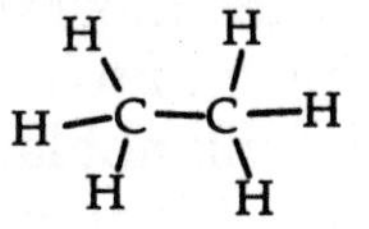

31.

a. vinyl chloride — H₂C=CHCl (structure shown)

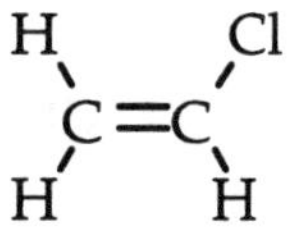

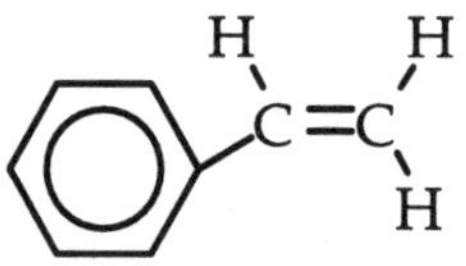

b. styrene — C₆H₅-CH=CH₂ (structure shown)

c. propylene — CH₂=CH-CH₃ (structure shown)

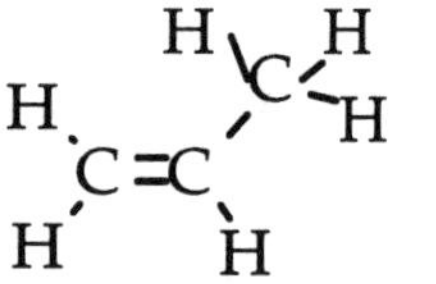

33. a. Polyester is a polymer built from ester monomers.
b. Polyamide is a polymer built from amide monomers.
c. Nylon 66 is a polymer built from 6-carbon monomers. The condensation of hexamethylene diamine and adipic acid yields a polymer with 6 carbon atoms between the nitrogen atoms.

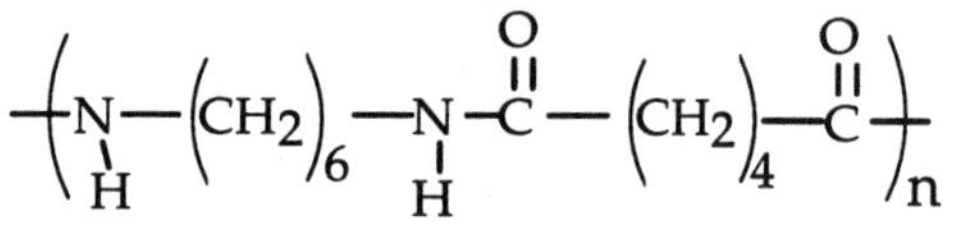

$$\left[-N(H)-(CH_2)_6-N(H)-\overset{O}{\underset{\|}{C}}-(CH_2)_4-\overset{O}{\underset{\|}{C}}- \right]_n$$

d. A diacid is a molecule containing two carboxylic acid groups.

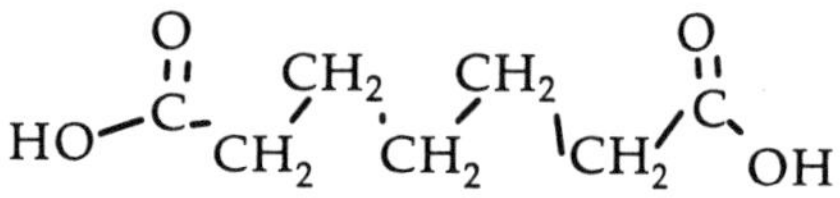

HO-CO-(CH₂)₅-CO-OH (structure shown)

e. A diamine is a molecule containing two amine groups.

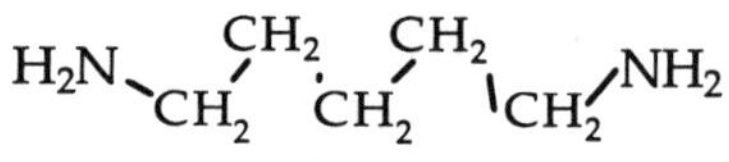

H₂N-(CH₂)₆-NH₂ (structure shown)

f. A peptide linkage has a carbonyl group, $\overset{\diagdown}{\underset{\diagup}{C}}=O$, adjacent to an amine -NHR. group.

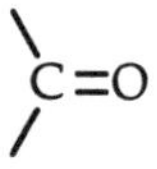

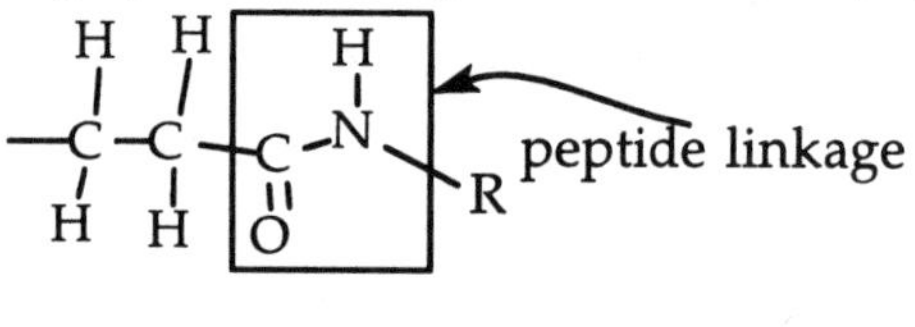

(structure shown with peptide linkage labeled)

35. All condensation reactions occur between two molecules to produce a larger molecule from the smaller reactant molecules. A separate small molecule like water which is eliminated.

37. Petroleum is the main source of the monomers to make most of todays polymers. Yes, this will change because petroleum reserves are declining and will not be as available in the future. Petroleum will become more expensive and will not last indefinitely. Plant material sources will be relatively cheaper in the future so they will probably become the major source of monomers for making polymers.

39. The four parts to successful recycling are: collection, sorting, reclamation, end-use. Unfortunately, collected materials will not be sorted or reclaimed if the supporting

infrastructure for processing doesn't exist. There must be an end use for reprocessed materials or else the collected materials will simply pile up somewhere.

41.
  a. aldehyde, alcohol
  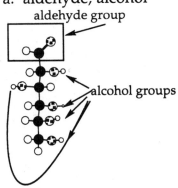

  b. alcohol, aromatic, amine

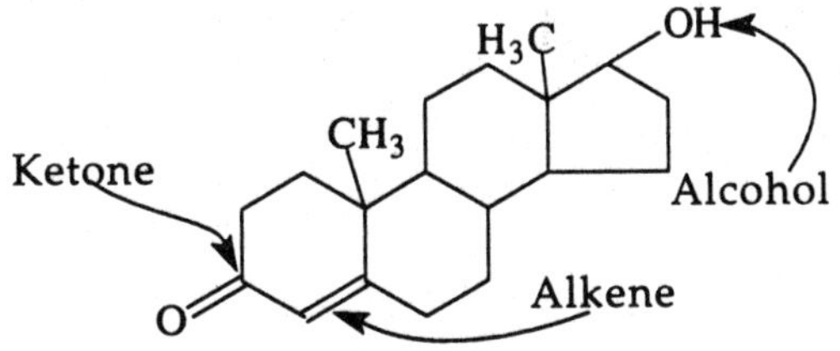

  c. aromatic, aldehyde, ether, alcohol

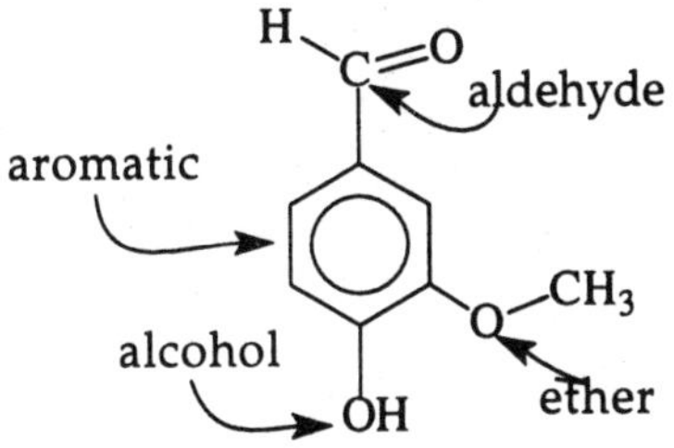

  d. alcohol, carboxylic acid

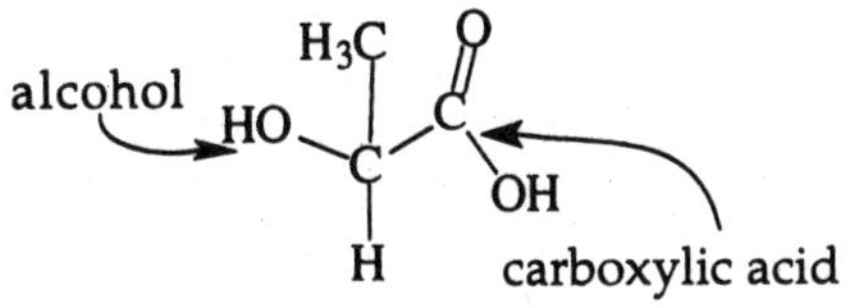

43. Polymer composite materials are substances that have a polymer matrix with reinforcing fibers. Glass-fiber-reinforced polyesters are used for boat hulls and car body panels. Graphite fiber-epoxy composites are used in fishing rods and tennis racquets.

Solutions for Problems

1. The number of monomer units in a polymer is determined by dividing the molar mass of the polyethylene polymer by the molar mass of the ethylene monomer. Determine the molar mass for ethylene. One mol CH_2CH_2 = 28 grams CH_2CH_2.

$$\frac{280{,}000 \text{ grams}}{1 \text{ mol polymer}} \times \frac{1 \text{ mol ethylene monomer}}{28 \text{ grams}} = \frac{100{,}000 \text{ ethylene monomers}}{1 \text{ polyethylene polymer}}$$

2. Propene, CH_3CHCH_2, has a molar mass of 42 g. 1 mol monomer = 42 g monomer. Divide the total molar mass for the polymer by the molar mass for the monomer. This gives the number of monomer molar masses. The molar mass of the polymer is 1 mol polymer = 84,000 g polymer
The number of monomer units

$$\left[\frac{84{,}000 \text{ grams polymer}}{1 \text{ mol polymer}}\right]\left[\frac{1 \text{ mol monomer}}{42 \text{ grams monomer}}\right] = 2{,}000 \text{ monomer units}$$

Speaking of Chemistry

Name _____

Organic Chemicals and Polymers

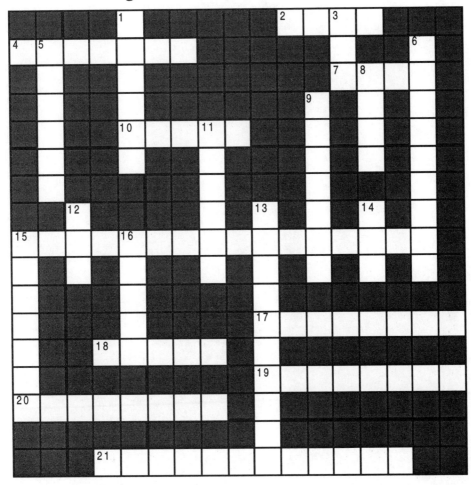

Across
2 High density polyethylene
4 Alcohol are ____, secondary and tertiary
7 Cross linked polyethylene
10 Made from dialcohol and diacid
15 _____ catalysts favor formation of one geometric isomer
17 Functional group in CH$_3$CHO
18 Twice the percent alcohol by volume
19 Monomer in natural rubber
20 Monomer of polyethylene
21 simplest aldehyde

Down
1 Odor produced by 3-methylbutyl acetate
3 Poly(vinyl chloride)
5 Poly-cis-isoprene
6 Codes are stamped on plastics to aid in _____
8 Low density polyethylene
9 Group that has an unpaired electron
11 Found in alcohols, ketones, esters, ethers
12 Poly(ethylene terephthalate)
13 _____ rubber is more elastic than natural latex
14 Alkene with substituents on the same side of double bond.
15 Monomer in polystyrene
16 Compounds like CH$_3$OCH$_3$

Bridging the Gap I   Name _____
Plastic Recycling: Classification and Sorting

Plastic recycling is legislated in many jurisdictions. The successful recycling of plastics or any material requires the four step recycling process of collection, sorting, reclamation, and end-use. A major problem with recycling plastics is that each type of plastic needs to be kept separate from the others to ensure a material that has an end-use market. If waste plastics are mixed, the reprocesser will not be dealing with a predictable polymer. The properties of reprocessed polymers can only be guaranteed by sorting the types of plastics. Right now consumers are asked to separate plastics and recycle using the codes shown here. Some recycling agencies even go so far as to say that if you cannot classify the material with certainty, you should not guess and possibly contaminate a batch of good recyclable materials. This exercise is aimed at giving you the practical experience of identifying plastic products you have in your surroundings. Your task is to look at the recycling codes stamped on plastic items in your home, apartment or even on the shelves of stores. You are to identify the plastic and note what initial product the polymer was used to make. You may have trouble finding some polymers, but you should try to identify objects for each polymer.

Plastic code and polymer name		Kind of plastic item (toy, bottle, tool, etc.)	Kind of plastic item (toy, bottle, tool, etc.)
1 PETE	polyethylene terephthalate		
2 HDPE	high density polyethylene		
3 V	polyvinyl chloride (PVC)		
4 LDPE	low density polyethylene		
5 PP	polypropylene		
6 PS	polystyrene		
7 OTHER	other resins and multilayered multi-material		

Is it reasonable that some polymers are more difficult to find? Why? Give your answers on back of this page.

© 1999 Harcourt Brace & Company. All rights reserved.

11 The Chemistry of Life

11.1 Handedness and Optical Isomerism

This section defines chiral, achiral, asymmetric carbons, and enantiomers. Examples of enantiomers are illustrated. Plane-polarized light, optical isomers and optical activity are described. The first separation of tartaric acid enantiomers by Louis Pasteur in 1848 is described. Dextro and levo terms for isomers are explained. The preference for L-isomers in natural systems is described. Racemic mixtures of equal amounts of enantiomers are described. Common examples of chiral compounds are identified.

Objectives
After studying this section a student should be able to:
- state the definition for chiral molecules and give examples
- give the definitions for achiral molecules and enantiomers
- inspect a structural formula and identify chiral atoms
- describe how plane-polarized light interacts with chiral molecules
- describe the work Pasteur did with tartaric acid
- give definitions for levo, L-isomers, and dextro, D-isomers
- tell what type of chiral molecules are preferred in nature
- describe why natural systems prefer L-isomers and do not usually use D-isomers

Key terms

handedness	nonsuperimposable	superimposable
achiral	chiral and chiral atom	enantiomers
asymmetric carbon atom	tetrahedral	lactic acid
dextro-, D-form	plane-polarized light	polarimeter
levo-, L-form	optical isomers	optically active
racemic mixture	monochromatic light	stereochemistry
chiral drugs	thalidomide	aspartame, NutraSweet

11.2 Carbohydrates

This section gives the general formula for carbohydrates, $C_x(H_2O)_y$, where x and y are whole numbers. Definitions and examples are given for carbohydrates, monosaccharides, disaccharides, and polysaccharides. Formulas are given for sucrose, glucose, lactose, D-glucose and D-galactose. The structures for the alcohol, aldehyde and ketone functional groups are reviewed. Hydrolysis reactions of disaccharides are illustrated. Diabetes types I and II are defined. The artificial sweeteners aspartame and saccharin are described. The structures and properties of polysaccharides, glycogen, cellulose, and starch are discussed.

Objectives
After studying this section a student should be able to:
- give definitions for carbohydrates, monosaccharides, disaccharides, and polysaccharides
- tell how disaccharides are formed from monosaccharides
- describe how disaccharides are converted to monosaccharides
- classify structures as mono-, di-, and poly- saccharides
- give reasons for using artificial sweeteners
- tell how starches, glycogen, and cellulose differ

Key terms

carbohydrate
disaccharides
dynamic equilibrium
sucrose
invert sugar

cellulose
polysaccharides
lactose
$C_{11}H_{22}O_{11}$
aspartame & saccharin

monosaccharides
hydrolysis
maltose
invertase
amylose

11.3 Lipids

Lipids are defined as waxes, fats, oils, and steroids. The general equation for the formation of fats (triglycerides) from fatty acids and glycerol is illustrated. Formulas for common fatty acids in oils and fats are given. The reasons for hydrogenation of vegetable oils are explained. The health concern about partially hydrogenated *cis* and *trans* unsaturated fatty acids is discussed. Monounsaturated and polyunsaturated fatty acids are illustrated. Linoleic acid and prostaglandins are discussed. The general structure for steroids is illustrated. Cholesterol and sex hormones are described.

Objectives

After studying this section a student should be able to:
- give definitions for lipids, waxes, fats, oils, steroids
- give definitions for monounsaturated, saturated, and polyunsaturated fatty acids
- if given a structural formula, classify the compound as a fat, fatty acid, steroid, or wax
- tell how *cis* and *trans* unsaturated fatty acids differ
- distinguish between a fat (ester) and a fatty acid
- tell how a partially hydrogenated oil differs from an unsaturated oil
- describe the function of prostaglandins
- describe the problem that cholesterol poses

Key terms

fats and oils
prostaglandins
solid triglycerides
steroids
cis fatty acids
waxes
estrogens

triglycerides
monounsaturated fats
PGE_2
liquid triglycerides
four-ring skeletal structure
trans fatty acids
progesterone

triester
essential fatty acid
PGE_α
hydrogenated
straight structures
cholesterol
testosterone

11.4 Soaps, Detergents and Shampoos

Definitions for surfactants, soaps and saponification are given. Example reactions illustrate the formation of a soap and glycerine from tristearin (an animal fat) and sodium hydroxide. The soap making process used by frontier farmers and pioneers is described. The cleaning action of soap is explained in terms of the molecular structure of soap and the formation of an emulsion.

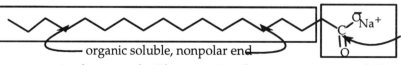

sodium stearate (a soap)

The emulsifying action is discussed. The reaction between soap and "hard" water ions (Ca^{2+}, Mg^{2+} and Fe^{2+}) to form soap "scum" or the proverbial bath tub ring is given.

This section describes the properties and structures of synthetic detergents. Anionic surfactants are defined and an example like the one below is illustrated. Similarly cationic surfactants are

```
[Oil-soluble part, hydrophobic] — [Water soluble part, hydrophilic: -O-S(=O)(=O)-O⁻ Na⁺]
```

described and illustrated. Nonionic detergents are defined and a typical structure is shown. The specific properties of each type of surfactant are described. Applications for each type of surfactant are given. Shampoos are described in terms of the surfactants and compounds needed to provide the desired properties. The structure and attributes of nonionic, anionic, and cationic surfactants are discussed. Representative structures for each class are illustrated. The chemical reasons why rinses solve "fly away hair" problems are given. The purpose of shampoo ingredients like lanolin and oil is explained.

Objectives

After studying this section a student should be able to:

- give the definition for surfactants
- describe the saponification process for making soap
- give definitions for "lye" , "soap" , " glycerin"
- sketch a soap molecule and identify the hydrophilic part and the hydrophobic part
- name the three metal ions that are found in "hard" water
- explain how soap scum forms
- explain how soap forms an emulsion when it dissolves oil
- tell how synthetic detergents differ from soap
- give a definition for hydrophobic and for hydrophilic
- sketch a detergent molecule and identify the hydrophobic and hydrophilic portions
- identify the negatively charged groups in anionic detergents
- sketch the general structure for an cationic detergent
- explain why a long hydrocarbon chain is not attracted to water molecules
- tell what happens when an anionic detergent mixes with a cationic detergent
- explain why ionic groups are attracted to water molecules
- give a definition for a shampoo
- name the type of detergent most frequently used to formulate a shampoo
- tell which one of the three types of detergents generally is a better foaming agent
- state the reason for adding nonionic surfactants to shampoos
- tell what if any connection there is between lather and cleaning efficiency
- tell why a cationic rinse is used after washing hair with an anionic shampoo
- explain why a cationic rinse makes damaged or disrupted hair feel smooth
- give two reasons for adding lanolin and mineral oils to shampoo

Key terms

surface tension	surfactants	soap
fat	alkali	saponification
salted out	hydrocarbon chain	nonpolar
hydrated	emulsion	hard water
hydrophobic	hydrophilic	anionic surfactants
oil soluble	quaternary ammonium halides	nonionic
cationic surfactants	cationic detergent rinse	surfactants
shampoos	anionic detergent	

© 1999 Harcourt Brace & Company. All rights reserved.

11.5 Creams and Lotions

The reasons for treating dry skin with oily substances are explained. Skin moisturizers and emollients are defined. The function of creams, lotions, emulsifying agents, and emulsions are explained. Colloids, oil-in-water emulsions, and water-in-oil emulsions are discussed. Cold cream is defined and barrier creams are described.

Objectives
After studying this section a student should be able to:
- give definitions for emollients and moisturizer
- name the two types of mixtures used as emollients
- give the definition for an emulsion
- explain the function of a barrier cream
- tell what a colloid is and give an example
- tell how to classify an emulsion as oil-in-water or water-in-oil
- explain how the cosmetic "cold cream" gets it name
- explain how the "dispersed phase" differs from the "continuous phase"

Key terms
cold cream	oils	moisturizer
creams lotions	emollients	emulsions
emulsifying agent	oil-in-water	water-in-oil
colloid	dispersed phase	continuous phase
barrier cream	shelf life	settle out

11.6 Amino Acids

This section gives definitions for and examples of amino acids. Essential amino acids are defined. Formulas for 20 common amino acids are tabulated.

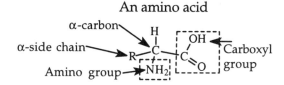

Objectives
After studying this section a student should be able to:
- draw and label the general formula for an amino acid
- tell how amino acids and proteins are related
- tell whether or not a molecule is an amino acid by looking at its structural formula
- give the definition for essential amino acids
- differentiate among amino acids R groups: basic, acidic, polar, and nonpolar

Key terms
amino group	amino acids	L-enantiomer
alpha carbon	carboxylic acid group	R group
essential amino acids	α-side chain	nonpolar
acidic	basic	polar

11.7 Peptides and Proteins

Formation of peptide bonds between amino acids is explained. The terms peptide, dipeptide, and polypeptide are defined. Simple proteins are illustrated using insulin as an example. Conjugated proteins are defined.

Objectives
After studying this section a student should be able to:
- draw the general structure for a peptide bond
- give a definition for a peptide linkage

give the definition for a dipeptide
draw the peptide structure formed by a specific acid and amine
state the definition of an amino acid residue
give a definition for simple proteins and name an example
give a definition for conjugated proteins
tell why proteins are called condensation polymers

Key terms

amide group	peptide	dipeptide
C-terminal end	"-ine" ending	-"yl" ending
amino acid residues	tripeptide	polypeptide
proteins	simple proteins	conjugated proteins
insulin	chymotrypsin	hemoglobin
myoglobin	condensation polymer	

11.8 Protein Structure and Function

This section shows how amino acid sequences can be indicated using abbreviations. The n! formula is explained using the 6 possible combinations for the tripeptides formed by Gly-Ala-Ser.

Gly-Ala-Ser	Ser-Gly-Ala	Ala-Gly-Ser
Gly-Ser-Ala	Ser-Ala-Gly	Ala-Ser-Gly

Neurotransmitters, enkephalins, and β-endorphins are described. The section defines the four levels of protein structure: primary, secondary, tertiary, and quaternary. Enzymes, active sites, and substrates are defined and enzyme action is explained.

Objectives

After studying this section a student should be able to:
give the definition for an amino acid sequence
tell the number of possible arrangements that can be formed by combining "n" amino acids
tell what the functions are for enkephalins and endorphins
give a definition for neurotransmitters
identify the amino acid sequence that enkephalins and β-endorphins have in common:
   the amino acid sequence Try-Gly-Gly-Phe
use the amino acid sequence such as Ala-Gly-Ser using the structures in Table 11.5 to
   draw the tripeptide structure and give its name
tell what determines the primary structure of proteins
give a definition for the secondary structure for proteins
name the two most common secondary structures
explain what makes up the tertiary structure of a protein
tell what denaturation does to a protein, and tell what can denature a protein
describe the function of enzymes
give definitions for substrate and active site
explain what genetic diseases are

Key terms

amino acid sequence	number of arrangements	n!
natural opiates	enkephalins	endorphins
neuropeptides	receptor sites	β-endorphins
neurotransmitters	enzymes	hormones
intermolecular	intramolecular	primary structure
secondary structure	tertiary structure	quaternary structure

β-pleated sheet disulfide bridge bonds α-helix
left-handed helices right-handed super helix globular proteins
globular enzyme denaturation energy of activation
active sites catalysts substrate
hydrolysis reaction condensation reactions reversible reaction
genetic diseases lactose intolerance

11.9 Hair Protein and Permanent Waves

This section describes the ionic and hydrogen bonds that exist between hair protein strands. The effects of water and pH on the bonds in hair are described. The chemical changes that occur in the permanent wave process are illustrated.

Objectives
After studying this section a student should be able to:
 describe the ionic bonds that exist between adjacent protein strands
 describe the effect of water on ionic bonds between proteins
 tell how hydrogen bonds are formed between adjacent protein strands
 describe the effect of water on hydrogen bonds between protein strands
 explain what effect ammonium thioglycolic acid has on disulfide bonds in hair
 tell why an oxidizer is needed in the permanent wave process

Key terms
ionic bonds sulfide linkages hydrogen bonds
disulfide cross-links static electrical charge pH
neutralizer permanent wave reducing agent

11.10 Energy and Biochemical Systems

This section reviews photosynthesis and oxidation of glucose as energy sources. The structures for ATP and ADP are given. The roles of ATP and ADP in energy transfer are described.

Objectives
After studying this section a student should be able to:
 describe the relationship between energy release and hydrolysis of ATP to form ADP
 identify the structures for ATP and ADP
 write the overall reaction for photosynthesis

Key terms
adenosine triphosphate ATP adenosine diphosphate
ADP photosynthesis

11.11 Nucleic Acids

This section gives the structures for the sugars, α-D-ribose and α-2-deoxy-D-ribose. The structures are illustrated for the five bases adenine(A), guanine(G), thymine(T), cytosine (C), and uracil(U) that make up nucleic acids. Nucleotides and trinucleotides are described. A trinucleotide sequence is shown. Complementary hydrogen bonding is discussed and illustrated. Protein synthesis is described in terms of mRNA, rRNA, and tRNA. Messenger RNA codons are tabulated for amino acids. Anticodons are defined and illustrated.

This section describes the process of using *E. coli* as gene factories. It illustrates the gene insertion process and gene recombination using tobacco mosaic virus and *agrobacterium tumefaciens* for altering tomato plant chromosomes. Examples of biogenetic engineering such as the production of transgenic animals and the biosynthesis of insulin are described.

The National Institute of Health (NIH) Human Genome Project to determine the complete sequence of base pairs in the human genome is described. Implications for diagnosing and treating hereditary diseases are discussed.

Objectives

After studying this section a student should be able to:
- give definitions for nucleic acids, DNA, and RNA
- name the five organic bases that form the message code in DNA and RNA
- name the three organic bases found in both DNA and RNA
- identify the structural components that are common to both RNA and DNA
- tell how the structures for RNA and DNA differ
- give definitions for nucleoside, nucleotide, and polynucleotides
- give a definition for complementary hydrogen bonding
- give definitions for gene and genome
- describe the process of protein synthesis
- if given a base sequence in DNA, tell what the base sequence is in complementary RNA
- use a mRNA code table to tell what amino acids a codon series dictates
- describe the process of gene splicing
- state reasons for and purpose of gene cloning
- explain why bacteria are used to produce recombinant DNA
- give reasons for developing transgenic species and identify two areas of research
- give an example of a product from transgenic "animals"
- describe the Human Genome Project
- tell why the NIH is carrying out the human genome project

Key terms

deoxyribonucleic acid	nucleic acids	DNA and RNA
ribonucleic acid	organic bases	α-2-deoxy-D-ribose
α-D-ribose	adenine(A)	thymine(T)
cystosine(C)	guanine(G)	uracil(U)
ribosomal RNA	messenger RNA	transfer RNA
polynucleotide	nucleotides	nucleoside
genes	trinucleotide	complementary bases
template	genome	complementary hydrogen bonding
transcription	replication	noncoding sequences
translation	codon	anticodon
spliced gene	recombinant DNA	base pair sequence
transgenic	transgenic animals	biogenetic engineering
human genome	NIH	tissue-plasminogen-activator

© 1999 Harcourt Brace & Company. All rights reserved.

Additional readings

Darnell Jr., James E. "RNA" <u>Scientific American</u> Oct. 1985: 68

Doolittle, Russell F. "Proteins" <u>Scientific American</u> Oct. 1985: 88.

Govindjee, Rajni. "The Absorption Of Light In Photosynthesis" <u>Scientific American</u> Dec. 1974: 68.

Hegstrom, Roger A. and Dilip K. Kondepudi. "The Handedness Of The Universe" <u>Scientific American</u> Jan. 1990: 108.

Kliener, Kurt. "Squabbling all the Way to the Genebank" (debate on accessibility of genetic information) <u>New Scientist</u> Nov. 26, 1994: 14.

McKenna, K. W., and V. Pantic, eds. "Hormonally Active Brain Peptides: Structure and Function" New York: Plenum Press, 1986.

Marshall, Eliot. "The Company that Genome Researchers Love to Hate (Human Genome Sciences Inc.)" <u>Science</u> Dec. 16, 1994: 1800.

Answers for Odd Numbered Questions for Review and Thought

1. Chiral is derived from the Greek word "cheir" for "hand". It is used to describe objects that are non-superimposable mirror images of one another. A pair of shoes is an example of two chiral objects.

3. A racemic mixture is a 50-50 mixture of enantiomers. The amount of optical activity produced by one enantiomer is canceled out by the activity of the mirror image.

5. An artificial sweetener is a nonsugar molecule that stimulates the sweetness receptors in human taste buds. Saccharin, cyclamate, and aspartame are examples of artificial sweeteners.

saccharin   cyclamate   aspartame

7. A lipid is a naturally occurring organic substance in living systems that is soluble in nonpolar organic solvents, but insoluble in water. The definition has nothing to do with structures or functional groups. This makes lipids a very odd lot of compounds. Waxes, fats, oils, and steroids are all water insoluble but very different structurally.

9. Hydrogenation is the conversion of C=C double bonds to single bonds by adding hydrogen atoms to carbon atoms in the molecule. The reaction produces a saturated hydrocarbon.

$CH_3CH_2CH_2CH_2CH_2CHCHCH_2CH_2CH_3 + H_2 \longrightarrow CH_3CH_2CH_2CH_2CH_2CH_2CH_2CH_2CH_2CH_3$

11. A soap or detergent molecule has one end that is ionic or polar covalent in nature; its other end is a nonpolar hydrocarbon. The soap or detergent molecule is long enough that the ends act independently. The charged end of the soap and detergent molecule is hydrophilic (attracted to water). The hydrocarbon end of the molecule is attracted to nonpolar solutes but not to water (hydrophobic). In use a layer of soap or detergent molecules forms around the greasy solute with hydrophilic groups radiating outward and with the hydrophobic hydrocarbon ends toward the greasy solute. This envelops the solute particles, makes them repel each other, and gives an outer layer that is attracted to the water so the solute particles "float away" in the water.

13. Magnesium and calcium salts of soaps are insoluble and will form a scum or residue when soaps are used in hard water. More soap is needed for cleaning since some has to first react with the "hard" ions to take them out of the water. Nonionic detergents do not form insoluble calcium or magnesium salts; hence, less is needed and they are more effective.

15. Ingredients are named in decreasing order of abundance in the formulation. The listing of water after the names of the oils indicates that oils are the dominant part of the mixture. The mixture is a water-in-oil emulsion.

17. The peptide bond is the bond formed between the amino group of one amino acid and the carboxyl carbon in another amino acid.

19. An amino acid contains the functional groups carboxyl, -COOH, and amino, $-NH_2$.

21. The "amino acid sequence" is the order of amino acid residues in a protein or peptide. An example is

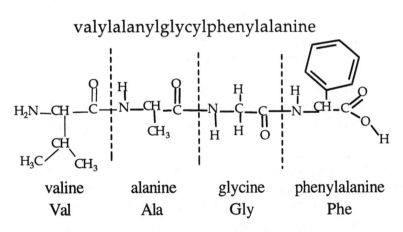

23. The protein primary structure is the sequence of amino acid residues.
    The secondary structure of proteins is usually either the regular repetitive pattern of the α-helix or the β-pleated sheet. Both are stabilized by hydrogen bonding. The tertiary

structure is the three dimensional structure resulting from the folding of the protein molecule. The quaternary structure refers to the shape formed by all the chains in the protein.

25. a. The α-helix is the spiral secondary protein structure in which the chain makes a complete turn every 3.6 amino acid residues. The coil is stabilized by hydrogen bonds between the oxygen of a $>C=O$ group and the hydrogen on an amino acid residue along the chain.
   b. The β-pleated sheet is a common secondary protein structure resulting from side to side intermolecular hydrogen bonding of extended polypeptide chains. The protein chains form parallel strands held together by hydrogen bonds between the H in an $-N{<}^H_H$ amino group and the O in the $>C=O$ group.

27. An active site is a part or region of an enzyme where the substrate sits down to interact with the enzyme.

29. a. The number of combinations can be predicted using the formula, n!, where n equals the number of amino acids. This means that for a tetrapeptide there are, 4 x 3 x 2 = 24, possible combinations.
   b. The 24 possible combinations for the amino acids glycine, Gly; alanine, Ala; serine, Ser: and cystine, Cys, are shown below. The combinations are determined by starting with one sequence and then systematically interchanging the positions of the amino acids. The table below was built up using this method.

Gly-Ala-Ser-Cys	Ala-Ser-Cys-Gly	Ser-Cys-Gly-Ala	Cys-Gly-Ala-Ser
Gly-Ala-Cys-Ser	Ala-Ser-Gly-Cys	Ser-Cys-Ala-Gly	Cys-Gly-Ser-Ala
Gly-Ser-Cys-Ala	Ala-Cys-Gly-Ser	Ser-Gly-Ala-Cys	Cys-Ala-Ser-Gly
Gly-Ser-Ala-Cys	Ala-Cys-Ser-Gly	Ser-Gly-Cys-Ala	Cys-Ala-Gly-Ser
Gly-Cys-Ala-Ser	Ala-Gly-Ser-Cys	Ser-Ala-Gly-Cys	Cys-Ser-Ala-Gly
Gly-Cys-Ser-Ala	Ala-Gly-Cys-Ser	Ser-Ala-Cys-Gly	Cys-Ser-Gly-Ala

31. The amino acids that are the in the tripeptide are valine, cysteine, tyrosine. The name for the tripeptide is Valylcystyltyrosine.

33. Enzymes are stereo specific they will only act on specific substrates. The D-glucose fits its active site in an enzyme, but the L-glucose will not act as a substrate for the enzyme. The human body lacks an enzyme that will act on L-glucose.

35. Protein chains in hair are held together by hydrogen bonds, ionic bonds and disulfide bonds. Hydrogen bonds can form between H atoms that are attached to very electronegative atoms (N and O) and the electron rich atoms with lone pairs such as N and O. The ionic bonds result when carboxylic acids (RCOOH) lose a proton to form carboxylate ions (RCOO⁻) and the basic amino groups (-NH$_2$) gain a proton to form -NH$_3^+$; these oppositely charged structures are attracted to one another. Disulfide bonds form between sulfur atoms in cysteine fragments in adjacent strands and the strands are held together by disulfide bonds, -S-S-, between cystine residues.

37. The disulfide bond is a single bond between two sulfur atoms. The disulfide bonds between cysteine amino acid units hold parallel strands of hair protein in place.

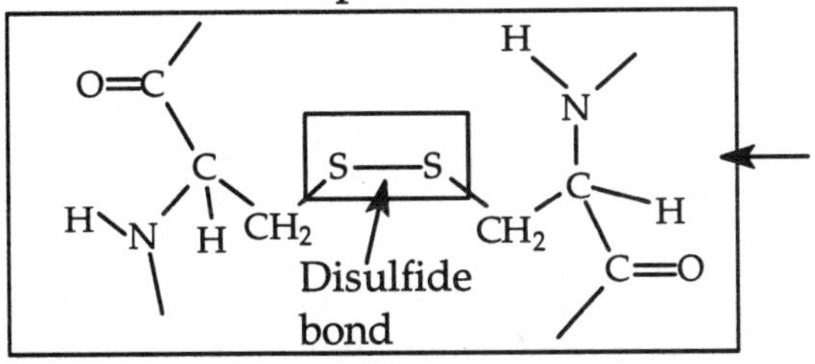

39. DNA and RNA differ structurally because each contains a different sugar for the polymer chain and they contain different bases. DNA contains the sugar, α-2-deoxy-D-ribose, $C_5H_{10}O_4$, while RNA contains α-2-D-ribose, $C_5H_{10}O_5$.

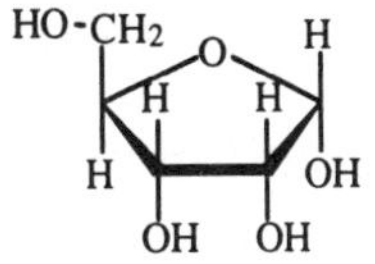

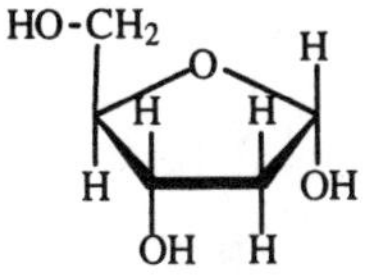

The base thymine only occurs in DNA. 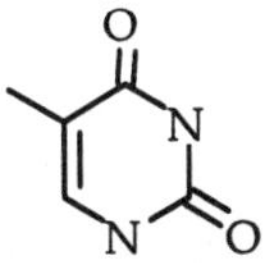   The base uracil only occurs in RNA.

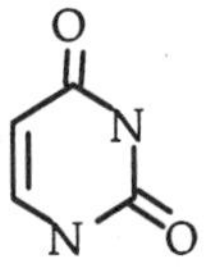

DNA and RNA are alike in that both DNA and RNA contain adenine, guanine, and cytosine.

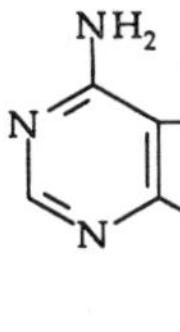

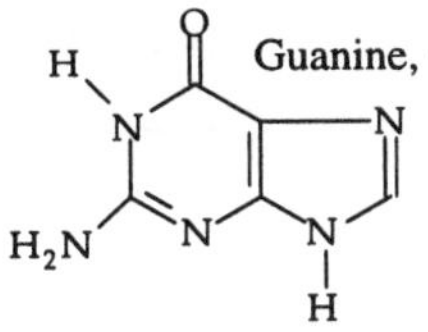

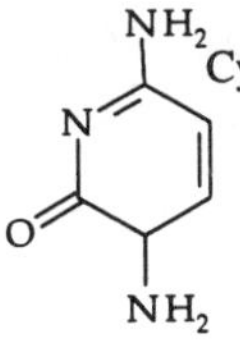

41. The monomers that polymerize to form DNA and RNA are nucleotides. The nucleotide monomers contain a phosphoric acid unit, a ribose unit and a nitrogenous base. See Figure 11.19 in text.

43. Complementary bases are pairs of bases like adenine-thymine and guanine-cytosine. These pairs have two hydrogen bond sites between the A-T pair and three hydrogen bonds between the G-C pair.

45. The human genome project is intended to completely map the sequence of base pairs in human DNA. This information may lead to the identification and treatment of hereditary diseases and diseases that may be triggered by changes in DNA function with aging.

47. The base pairs between DNA and mRNA are

DNA	mRNA
G, guanine	C, cytosine
A, adenine	U, uracil
C, cytosine	G, guanine
T, thymine	A, adenine

This means the best way to determine the sequence of bases in a complementary mRNA strand is to write out the DNA sequence and then one by one write the matching base immediately below.

a. The DNA strand with the base sequence T G T C A G T G G G C C G C T has a complementary mRNA sequence of A C A G U C A C C C G G C G A. The pairings look like this

 DNA sequence   T G T C A G T G G G C C G C T
 mRNA sequence A C A G U C A C C C G G C G A.

The complementary pairings of bases between RNA strands are
 mRNA   tRNA
 G, guanine C, cytosine
 A, adenine U, uracil

The tRNA anticodon order can be determined by writing out the mRNA sequence and then translating the complementary pairs one by one. There are essentially only two pairs to deal with so translation is not complicated.

b. Here the pairings would look like this.
 mRNA sequence A C A G U C A C C C G G C G A
 tRNA sequence   U G U C A G U G G G C C G C U

The anticodon order in the tRNA would be U G U C A G U G G G C C G C U

c. The mRNA codes for amino acids are tabulated in Table 11.6.
The table shows that the sequence of fifteen bases contains only five codons
U G U C A G U G G G C C G C U
[U G U][C A G][U G G][G C C][G C U] The amino acids that match these codons are:
 U G U for Cysteine C A G for Glutamine U G G for Tryptophan
 G C C for Alanine   G C U for Alanine.

The amino acid sequence is Cysteine Glutamine Tryptophan Alanine Alanine.

49. Watson and Crick proposed that the double helix was stabilized by hydrogen bonding between specific bases that had complementary hydrogen bonding. In these cases bases are attracted more to one another because of the match between sites in both members.

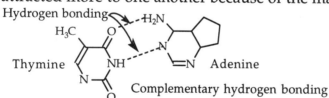

51. The DNA sequence   ... G T A G C ... has a complementary
  mRNA sequence   ... C A U C G ....

53. DNA sequence   mRNA sequence
  G   C
  A   U
  C   G
  A   U

Speaking of Chemistry

The Chemistry of Life

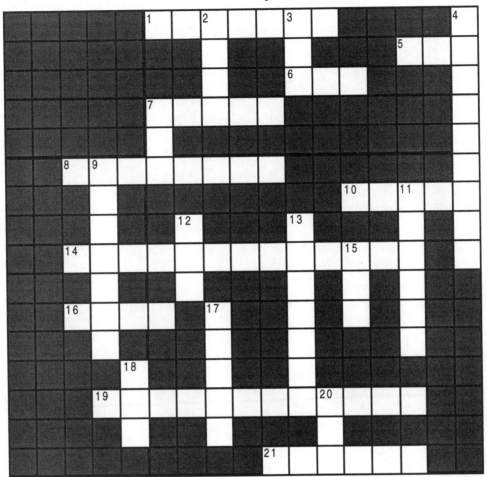

Across
1. Separated enantiomers of tartaric acid in 1848
5. Ribonucleic acid
6. Codon for start or methionine
7. carbon adjacent to acid functional group
8. Bond between atoms that share electrons equally
10. Unequal sharing of electrons
14. Carbohydrates like sucrose or maltose
16. Solid triglycercides
19. Oils can be made more solid if partially _____
21. Waxes, fats, oils, steroids

Down
2. Produced from lye and fats
3. Stop codon
4. Artificial sweetener
7. Adenosine triphosphate
9. Chiral molecules are ____ isomers
11. Essential amino acid
12. Serine codon
13. Amino acid polymer
15. Deoxyribonucleic acid
17. Three base code
18. NaOH
20. Adenosine diphosphate

© 1999 Harcourt Brace & Company. All rights reserved.

Bridging the Gap I
Tetrahedron Construction and Chiral Centers

Two dimensional illustrations in the chapter are excellent, but they cannot give a tactile sense of the tetrahedral form and the non superimposability of molecules that are mirror images of one another. Enantiomers are molecules that have the same molecular formula but different spatial arrangements. They are mirror images of one another. This exercise is intended to give you an experience with three dimensional models of the lactic acid enantiomers.

Please read all these directions before doing any cutting.

Write your name in the blank space provided on the templates. Your instructor may want you to turn in your completed tetrahedra. Be careful to keep the A, B, and D tabs on the template when you cut out the tetrahedron. Be sure to leave the black edges on the faces. Color code the four different groups at the corners of the tetrahedron before cutting. A suggested color code is given here.

Group	hydrogen, -H	hydroxyl, -OH	methyl, -CH3	carboxylic acid, -COOH
Color	white	blue	green	red

Hold the cutout so you can read your name. Fold faces A, B, and D away from you. Hold face D up so you can read it. Fold the hidden support away from you. Do this same process with face B and the second hidden support. Slide the hidden support behind face A. Fold tab B over the B on face B. Insert the remaining hidden support behind face B. Fold tab A over the A on face A. Fold tab D over the D on face D. All the tabs can be secured with a piece of transparent tape if you wish. You now have your tetrahedron that represents lactic acid.

The methyl group (-CH3), hydroxyl (-OH), carboxylic acid group (-COOH), and hydrogen (-H) are on the corners of the tetrahedron. A carbon atom is in the center of the tetrahedron. There are two different tetrahedral structures for lactic acid, one is the mirror image of the other. After constructing the two tetrahedra hold a small mirror next to one of the models. You should rotate the other model to match the appearance of the image in the mirror. You now can see why the pairs are called mirror images of one another. Put away the mirror and try to rotate the two models to see if you can match the positions of the colors (substituents). If you can, they are superimposable. If you cannot match them point for point at all points, then the models are nonsuperimposable.

The drawing below illustrates two isomers that are mirror images of one another.

© 1999 Harcourt Brace & Company. All rights reserved.

Chiral Tetrahedron #1

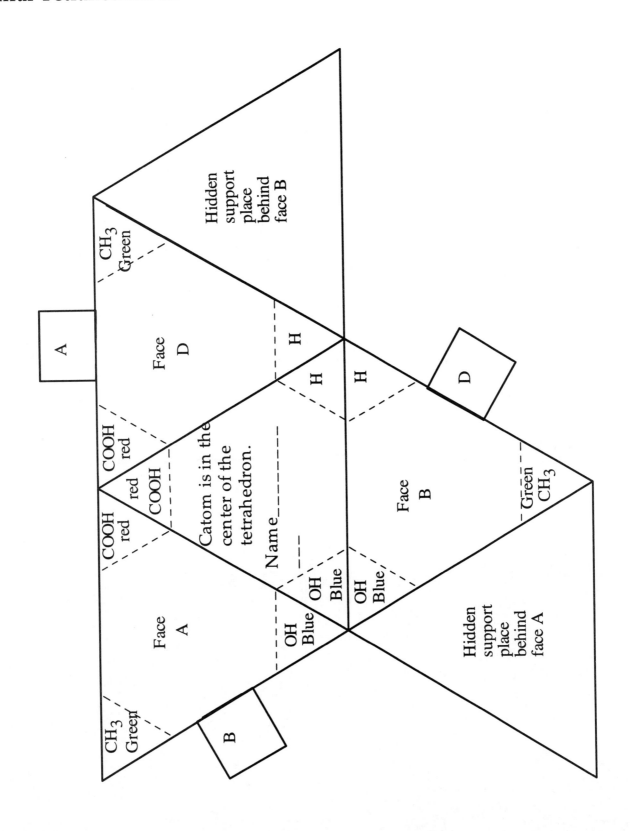

Chiral Tetrahedron #2

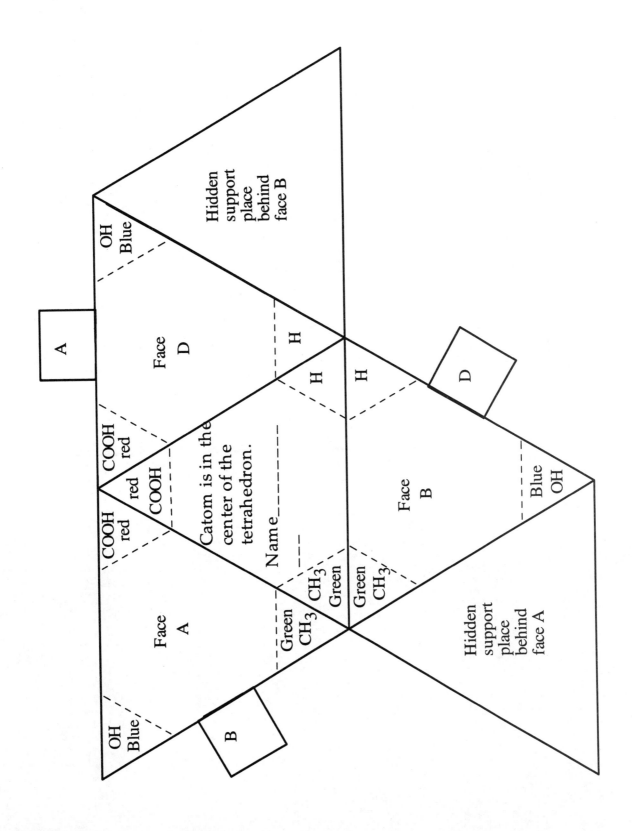

12 Nutrition: The Basis of Healthy Living

12.1 Digestion: It's Just Chemical Decomposition

The digestion process is described. The types of food digested in each part of the digestive system are identified. The conversions of carbohydrates to simple sugars, fats to glycerol, and proteins to amino acids are described. The fate of these small molecules formed by digestion is summarized.

Objectives
After studying this section a student should be able to:
- name the three major parts of the digestive system
- name the products formed by the digestion of carbohydrates, fats and proteins
- match a type of food with the organ where it is digested
- tell what products result from the digestion of carbohydrates, fats and proteins
- tell what happens to the fatty acids, amino acids and simple sugars formed by digestion

Key terms

digestion	mouth	amylase	carbohydrates
starch	stomach	small intestine	protein
lipids	energy production	biomolecules	storage as fat

12.2 Energy: Use It or Store It

The BMR is defined as the minimum energy needed by the body to maintain body temperature, support cell reactions, and control messages passing through the nervous system. The BMR is calculated according to this formula. During adulthood the BMR decreases by 2% -3% with each additional ten years of age.

$$\text{BMR in Calories/day} = 10 \times \text{body weight in pounds}$$

The method for calculating a person's minimum daily caloric needs is shown. Activity level factors are tabulated in Table 12.3.

$$\text{Calories needed/day} = \text{BMR} \times \text{activity level factor}$$

The energy available from a gram each nutrient is given. Glucose is the body's basic fuel molecule.

Fats: $\dfrac{9 \text{ kcal}}{1 \text{ gram}}$ Carbohydrates: $\dfrac{4 \text{ kcal}}{1 \text{ gram}}$ Proteins: $\dfrac{4 \text{ kcal}}{1 \text{ gram}}$

Energy expenditures for various activities are shown in Table 12.1. The body stores energy in ATP. This energy is released when ATP is converted to ADP. Carbohydrate, fat, protein and water content for some common foods are tabulated in Table 12.2.

Objectives
After studying this section a student should be able to:
- give a definition for basal metabolic rate (BMR)
- tell how to estimate a person's BMR using their body weight in pounds
- tell how a person's minimum daily caloric needs are estimated
- tell how many kcal are released by a gram of fat, carbohydrate, or protein
- calculate the minimum daily caloric need of a person from activity level factors and BMR
- calculate the number of kcal that can be obtained from a sample of food if given the grams of food and the percentages for fat, carbohydrate and protein
- tell what ATP and ADP do in the production of energy

© 1999 Harcourt Brace & Company. All rights reserved.

Key terms

    basal metabolic rate, BMR   Calorie   kilocalorie
    fats: 9 kcal/g   carbohydrates: 4 kcal/g   proteins: 4 kcal/g
    energy taken in   energy used   energy stored
    glucose   ATP

12.3 Sugar and Polysaccharides, Digestible and Indigestible

Carbohydrates are described in terms of digestible and indigestible. Digestible carbohydrates are simple sugars, disaccharide's, and polysaccharides (amylose, amylpectin, and glycogen). Specific indigestible carbohydrates are identified as cellulose and its derivatives, pectin and plant gums. Dietary fiber is defined and its importance is explained.

Objectives

After studying this section a student should be able to:
    name three kinds of digestible carbohydrates
    identify two indigestible carbohydrates
    name sources of insoluble fiber and soluble fiber
    describe how insoluble fiber influences risks of heart disease and colorectal cancer

Key terms

simple sugars	glucose	fructose
disaccharides	sucrose	maltose
polysaccharides	amylose	amylpectin
indigestible carbohydrates	cellulose	pectin
dietary fiber	insoluble fiber	soluble fiber
heart disease	colorectal cancer	

12.4 Lipids, Mostly Fats and Oils

Sources of dietary fats are identified. The structure and breakdown of stored body fat is discussed. Heart disease (atherosclerosis) is described. The link between heart disease and fats is explained. The public health authorities recommend following a diet containing 30% or less of fat. The formation of lipoproteins from cholesterol is described. The role of HDL and LDL in heart disease is explained.

Objectives

After studying this section a student should be able to:
    name the common sources of dietary fats
    tell what two types of compounds result from hydrolyzing fats
    tell why high fat intake increases the risk of heart disease
    give a description of heart disease
    describe the connection between cholesterol and lipoproteins
    give definitions for HDL and LDL
    tell how good and bad cholesterol differ relative to the risk of heart disease

Key terms

lipids	fats	triglycerides
saturated fat	glycerol	free fatty acids
atherosclerosis	monounsaturated fat	polyunsaturated fat
cholesterol	lipoproteins	LDLs
HDLs	"bad" cholesterol	"good" cholesterol
plaque	heart attack	

12.5 Proteins in the Diet

This section identifies high protein foods. The function of dietary protein is explained. The fate of excess dietary protein is described.

Objectives
After studying this section a student should be able to:
- identify the major dietary sources of protein
- describe the role of dietary protein
- tell what the body does with excess protein

Key terms
high-protein foods	amino acids	excess amino acids
urea	ammonia	

12.6 Our Daily Diet

An explanation of how to read and interpret a sample Nutrition Facts Label. Micronutrients and macronutrients are defined. The reference daily values are described. The mandatory listings in a Nutrition Facts Label are identified. Health risks associated with dietary deficiencies of iron, calcium, iodine, and other nutrients are identified.

Objectives
After studying this section a student should be able to:
- name the nutrients that are in the mandatory listings on the Nutrition Facts Label
- describe the basis of the "daily reference values" for nutrients
- explain why iron, calcium, vitamin A, and vitamin C are in the mandatory list
- tell what deficiencies cause osteoporosis and anemia
- describe the symptoms associated with osteoporosis and anemia

Key terms
% Daily Value	macronutrients	Nutrition Facts Label
daily reference values, DVs	mandatory listings	micronutrients
osteoporosis	cancer risk	anemia

12.7 Vitamins in the Diet

Vitamins are defined as organic compounds essential to health that must be supplied in small amounts in the diet.. The water-soluble and fat-soluble classes of vitamin are described. The structures and functions of specific vitamins are summarized. Typical sources of vitamins A, D, E, K, B, and C are given. Structures for riboflavin, thiamine and vitamin C are illustrated.

Objectives
After studying this section a student should be able to:
- give the definition for vitamins
- explain why some vitamins are water-soluble and others are fat soluble
- tell why the solubility classification of vitamins is important
- tell which types of vitamins are stored in the body and which are not
- tell what is meant by a vitamin megadose
- name some typical water-soluble vitamins and some typical fat-soluble vitamins
- describe the functions of vitamins such as A, D, E, K, B and C
- explain what an antioxidant does and name a vitamin that is an antioxidant

© 1999 Harcourt Brace & Company. All rights reserved.

give a definition for a coenzyme
name a health problem that results from a deficiency of each vitamin A, D, E, K, B and C

Key terms

vitamins	megadoses	fat-soluble
nonpolar	vitamins A, D, E, and K	antioxidant
night blindness	rickets	water-soluble
coenzymes	B vitamins	pellagra
vitamin C, ascorbic acid	scurvy	placebo

12.8 Minerals in the Diet

Mineral nutrients are defined as elements other than carbon, hydrogen, nitrogen, and oxygen. Macrominerals make up about 4% of body weight. Micronutrient minerals are essential in small amounts. Four (iron, copper, zinc, and iodine) are allowed to be listed on Nutrition Facts labels. Sodium, potassium, and chloride are essential in electrolyte balance. Sources of each mineral are tabulated. Minerals and their deficiency diseases are identified: calcium and osteoporosis; iodine and thyroid function; iron and anemia.

Objectives

After studying this section a student should be able to:

name sources for Na^+, K^+, Cl^-, Ca^{2+}

describe the role of each mineral: Na^+, K^+, Cl^-, Ca^{2+}

tell what mineral is needed to prevent osteoporosis

describe the role sodium ion concentration in high blood pressure

describe the health problems that result from deficiencies in iodine, iron, calcium & zinc

tell what mineral is typically deficient when a person has anemia

Key terms

minerals	macronutrient minerals	calcium
potassium	magnesium	sodium
electrolyte balance	chlorine	high blood pressure
osteoporosis	normal blood pressure	iron
copper	micronutrient minerals	iodine
thyroid	zinc	anemia

12.9 Food Additives

Food additives are added to protect the food from spoiling by oxidation, bacterial attack, or aging. The function and range of food additives are described. The Food and Drug Administration (FDA) GRAS list includes additives "generally recognized as safe". The specific functions of food additives are defined: preservatives, antioxidants, sequestrants, flavorings, flavor enhancers, food colorings, pH adjusters, anticaking agents, stabilizers and thickeners. Examples of each type of food additive and their benefits are given.

Objectives

After studying this section a student should be able to:

explain why food additives are necessary
tell what is meant by the GRAS list
explain why foods can be preserved by drying
describe the way that concentrated salt and sugar solutions preserve foods
tell why freezing helps preserve foods
name two antioxidants and tell why they are added to foods
tell why sequestrants are added to food

explain why food flavorings and flavoring enhancers are added to food

tell why BHA and BHT are added to food

tell why pH adjusters are added to food

Key terms

food additives	GRAS list	preservatives
drying	dehydrate	osmosis
sodium benzoate	sodium propionate	salted
antioxidants	rancid	BHT
BHA	sequestrants	complexes
food flavors	volatile oils	extracts
flavor enhancers	potentiators	MSG
food colors	pH control	buffers
preservatives	antioxidants	viscosity modifiers
anticaking agents	hygroscopic	magnesium silicate
stabilizers	thickeners	hydroxyl groups

12.10 Some Daily Diet Arithmetic

This section shows how to calculate the percent fat in a food using the information on a Nutrition Facts Label. An example calculation shows how to calculate the caloric value of a food from the grams of fat, grams of carbohydrate and grams of protein. An example calculation is given that shows how to determine the number of kcal and grams of fat, protein, and carbohydrate a person should consume to meet their daily calorie requirement and stay in the recommended percentages of fat, carbohydrates and protein.

Objectives

After studying this section a student should be able to:

read a Nutrition Facts Label to find the grams of fat, carbohydrate and protein in the food

read the label to find the total Calories and the Calories from fat and then calculate % Calories from fat

read a food label to find the grams of fat, protein and carbohydrate, calculate the total Calories, and the % Calories from fat

calculate the %Calories from fat if given total Calories and number of Calories from fat

calculate the Calories from fat, protein, and carbohydrates given total Calories and the % from fat, % from protein and % from carbohydrates

state the number of dietetic Calories available from 1 gram of fat, protein, carbohydrate

Key terms

Nutrition Facts labels total calories from fat

$$\% \text{ Calories from fat} = 100 \times \left(\frac{\text{Calories from fat}}{\text{Total calories}}\right)$$

$$\frac{9 \text{ kcal}}{\text{g fat}} \qquad \frac{4 \text{ kcal}}{\text{g protein}} \qquad \frac{4 \text{ kcal}}{\text{g carbohydrate}}$$

Additional readings

Heaney, Robert P. " Protein Intake and the Calcium Economy." Journal of the American Dietetic Association 22 Nov. 1993, 1259.

Hunter, Beatrice T. " How Safe are Nutritional Supplements?" Consumers' Research Magazine Mar. 1994: 21.

Hunter, Beatrice T. " Small but Significant. (role of trace elements in human health)" Consumers' Research Magazine Dec. 1994: 8.

Sangiorgio, Maureen et al. "A Salty Surprise. (sodium sources)" Prevention Feb. 1992: 12.

" Taking Vitamins: Can They Prevent Disease?" Consumer Reports Sept. 1994: 561.

Tarnopolsky, Mark. "Protein, Caffeine, and Sports: Guidelines for Active People" The Physician and Sportsmedicine March. 1993: 137.

"The Facts About Fats." Consumer Reports June 1995: 389.

Wurtman, Richard J. "Nutrients That Modify Brain Function" Scientific American April 1982: 50.

Yulsman, Tom. "HDL Finally gets a Hearing." Medical World News March 1992: v33 n3 27.

Answers to Odd Numbered Questions for Review and Thought

1. Digestion is the process of breaking large molecules in food into substances small enough to be absorbed by the body from the digestive tract.

3. The energy and mass relations for fats, proteins and carbohydrates are as follows:
   a. Fats: 9 kcal/g or 9 Calories/g
   b. Proteins: 4 kcal/g or 4 Calories/g
   c. Carbohydrate: 4 kcal/g or 4 Calories/g

5. Metabolic energy is the energy expended to keep the heart beating, the lungs inhaling and exhaling air, the nerves generating and transmitting their flood of electrical impulses, and the cells of the body conducting their normal functions. The basal metabolic rate, BMR, is the rate at which the body uses energy to support these maintenance operations. It is typically in kcal or Calories per hour. It is the minimum energy required for a person to stay alive.

7. A triglyceride is a triester formed from glycerol and three fatty acids. The three fatty acids do not have to be the same carbon chains. The fatty acids chain length ranges from 4 to about 24 carbons. Most of the naturally occurring fatty acids have even numbered chain lengths

9.
   a. Atherosclerosis is a disease that is associated with the buildup of fatty deposits on the inner walls of arteries, i.e. hardening of the arteries, notably in the brain and in the heart.. These deposits lead to higher risks of heart attacks.
   b. Cholesterol is a lipid steroidal alcohol that contributes to the development of artherosclerosis.
   c. Plaque is the yellowish deposit of cholesterol and lipid-containing material on artery walls that is a symptom of artherosclerosis. These deposits can promote blood clots that cut off blood supply and cause heart attacks and strokes.
   d. Lipoproteins are complex assemblages of lipids, cholesterol and proteins that serve to transport water-soluble lipids in the blood stream through the body.
   e. Low density lipoproteins are richer in lipid than in protein and have a corresponding lower density in the range of 1.006-1.063 g/mL.
   f. High density lipoproteins are richer in protein than in lipid with a corresponding higher density in the range of, 1.063-1.210 g/mL.

11.
   a. Macronutrients are nutrients the body needs in large amounts.
   b. Micronutrients are nutrients needed by the body in small amounts.

13.
   a. Starches come from plants; they are found in foods like pasta, rice, potatoes, corn.
   b. Glycogen is produced in the liver and limited amounts are stored in liver tissue and muscle.
   c. Cellulose comes from plants.

15. Vitamins are organic compounds essential to health; they are needed in small amounts in the diet. Vitamins are not synthesized by the body and must be obtained from plant sources. There are approximately 14 compounds identified as vitamins. Vitamins function as coenzymes.

17. An electrolyte is a substance that dissolves in water to produce ions. The electrolyte balance refers to a condition of proper transfer of material through osmosis, normal nerve impulse transmission, normal acid-base balance and extracellular volume.

19. The seven most abundant macronutrient minerals in the body are phosphorus, calcium, magnesium, sodium, potassium, chlorine, and sulfur.

21. Sequesterants are added to food to complex and tie up metal ions to keep them from catalyzing the decomposition of food.

23. An upper limit of 30% of daily calories are recommended to come from fat.

25. Milk should be promoted as a healthy food for the general public because it is an excellent source of calcium and protein. Milk should not be promoted as a healthy food because people who are lactose intolerant may experience serious adverse reactions and become ill when they consume milk.

27. Cholesterol is water insoluble because water is a small polar molecule that will have very weak attractions to the large, organic nonpolar cholesterol molecule. High levels of cholesterol are linked to atherosclerosis or hardening of the arteries. There are 27 carbon atoms in cholesterol, $C_{27}H_{45}OH$.

29. Women produce less estrogen after menopause. Estrogen inhibits calcium loss and bone erosion. Reduced estrogen levels result in more rapid calcium and bone loss. Osteoporosis can be minimized if adequate calcium intake occurs, especially from adolescence through young adulthood, and if estrogen replacements are taken after menopause. Exercise also helps. The benefit of exercise results from the fact that bone structure is more dense when the bones are subjected to loads associated with exercise.

31. An iodine deficiency leads to enlargement of the thyroid. This condition is called "goiter". Potassium iodide, KI, is added to table salt to make iodized salt in order to prevent iodine deficiency.

33. There is no right answer for this question. Each answer will depend on the specific product.

Solutions for Problems

1. In this case BMR = 10 x weight of 110 pounds = 1100 Calories. The activity factor is 1.6. The estimated daily caloric need is = 10 x 110 x 1.6 = 1100 x 1.6 = 1760 Calories.

2. The BMR is calculated by multiplying the person's body weight by a factor of 10. BMR = 145 x 10 = 1450   If he increases his activity, his BMR will increase.

3. Bicycling uses 5 Calories for each minute of riding time, or in 1 minute a person will consume 5 Calories. Using 100 Calories will require 20 minutes of riding time.

$$100 \text{ Calories} \times \frac{1 \text{ min}}{5 \text{ Cal}} = 20 \text{ minutes}$$

4. The key number here is the number of calories that must be burned off. The 207 Calories will be burned at a rate of 10 Calories every minute when running at 5 miles/hour. This comes from Table 12.1 on page 307.

$$\text{minutes} = \frac{207 \text{ Calories}}{1} \times \frac{1 \text{ kcal}}{1 \text{ Calorie}} \times \frac{1 \text{ minute}}{10 \text{ kcal}} = 20.7 \text{ minutes}$$

5. Walking at 3.5 miles/hour consumes 3.5 Calories per minute. The 100 grams of bread provides 269 Calories. This 269 Calories will be consumed in 77 minutes.

$$\text{minutes} = \frac{269 \text{ Calories}}{100 \text{ g bread}} \times \frac{1 \text{ minute walking}}{3.5 \text{ Calories}} = \frac{77 \text{ minutes}}{100 \text{ g bread}} \text{ or}$$

$$269 \text{ Calories} \times \frac{1 \text{ minute walking}}{3.5 \text{ Calories}} = 76.86 = 77 \text{ minutes}$$

6. The total blood volume in mL multiplied by the milligrams per 100 mL of blood will give the milligrams of cholesterol in the body.

$$\frac{200 \text{ mg cholesterol}}{100 \text{ mL blood}} \times \frac{12 \text{ pints blood}}{\text{body}} \times \frac{473 \text{ mL}}{1 \text{ pint}} = \frac{11352 \text{ mg cholesterol}}{\text{body}}$$

$$= \frac{11400 \text{ mg cholesterol}}{\text{body}}$$

7. Since we know that the dangerous cholesterol level is 240 mg per 100 mL of blood, we need to express the person's blood volume in milliliters. We need to change 13 pints to milliliters. Then we can multiply by the number of mg of cholesterol in 100 mL of blood to get the person's total amount.

$$13 \text{ pt} \times \frac{473 \text{ ml}}{1 \text{ pt}} \times \frac{240 \text{ mg cholesterol}}{100 \text{ ml blood}} = 14758 \text{ mg} = 15000 \text{ mg}$$

Convert the milligrams to grams, $15000 \text{ mg} \times \dfrac{1 \text{ gram}}{1000 \text{ milligrams}} = 15 \text{ grams}$

This is 15 grams. Round to 2 significant digits because 13 only has 2 sd.

8. It is easier to make the comparison if we determine the % of calories due to the fat in each piece of pie.

pecan pie: $\dfrac{180 \text{ Calories fat}}{459 \text{ Calories total}} \times 100 = 39.2$ % of calories from fat

apple pie: $\dfrac{108 \text{ Calories fat}}{305 \text{ Calories total}} \times 100 = 35.4$ % of calories from fat

The apple pie is a better low fat choice.

9. Total Calories are found by adding together the Calories from the protein, carbohydrate, and fat; protein and carbohydrate have 4 Cal. per 1 gram while fat has 9 Cal. per gram.

Calories total = Calories from protein + Calories from carbohydrate + Calories from fat

Calories from protein = $27 \text{ g} \times \dfrac{4 \text{ Cal}}{1 \text{ gram}} = 108$ Calories

Calories from carbohydrate = $46 \text{ g} \times \dfrac{4 \text{ Cal}}{1 \text{ gram}} = 184$ Calories

Calories from fat = $34 \text{ g} \times \dfrac{9 \text{ Cal}}{1 \text{ gram}} = 306$ Calories

108 + 184 + 306 = 598 Calories total

10. Total Calories are found by adding together the Calories from the protein, carbohydrate, and fat; protein and carbohydrate have 4 Cal. per 1 gram while fat has 9 Cal. per gram.

Calories from protein = $34 \text{ g} \times \dfrac{4 \text{ Cal}}{1 \text{ gram}} = 136$ Calories

Calories from carbohydrate = $44 \text{ g} \times \dfrac{4 \text{ Cal}}{1 \text{ gram}} = 176$ Calories

Calories from fat = $46 \text{ g} \times \dfrac{9 \text{ Cal}}{1 \text{ gram}} = 414$ Calories

136 + 176 + 414 = 726 Calories total

11. First calculate her BMR, BMR = 145 x 10 = 1450 Calories.

    Now calculate her daily Calorie need; since she is extremely active, her activity factor is 1.9.
    1450 Calories x 1.9 = 2755 Calories for her daily Calorie need.

    Now find 30% of 2755 Calorie to give the number of Calories to come from fat, 60% of 2755 Calorie to get the Calories from carbohydrates, and 10% of 2755 Calorie to get the Calories needed from protein. These should be rounded off to two significant digits because of the two digits in the activity factor of 1.9. Finally convert each number of Calories to grams of substance using $\frac{1 \text{ gram}}{4 \text{ Cal}}$ for carbohydrate and protein and $\frac{1 \text{ gram}}{9 \text{ Cal}}$ for fat.

    grams of fat = .30 x 2755 Cal x $\frac{1 \text{ gram}}{9 \text{ Cal}}$ = 92 g fat;

    grams of carbohydrate = .60 x 2755 Cal x $\frac{1 \text{ gram}}{4 \text{ Cal}}$ = 413 = 410 g carbohydrate;

    grams of protein = .1 x 2755 Cal x $\frac{1 \text{ gram}}{4 \text{ Cal}}$ = 69 g protein.

12. You determine the percent one serving provides of a person's total Calorie need by using the following:

    Percent = 100 x $\left[\frac{\text{Calories provided}}{\text{Total Calories needed}} \right]$

    The person's daily need is 1800 Calorie. The label says one serving provides 100 Cal.

    % of daily Calorie need in the serving = $\frac{100 \text{ Calories serving}}{1800 \text{ Calorie daily need}}$ x 100 = 5.55 = 5.6 %

    % of Calories from fat in the serving = $\frac{20 \text{ Calories fat}}{100 \text{ Calorie serving}}$ x 100 = 20 %

    This food would not be a good product for someone on a low salt (low sodium, Na^+) diet. This product contains 730 mg sodium or 30% of someone's daily amount in each serving. Getting 30% of your daily sodium in only 5.6% of your daily Calories is not a good combination. The recommended total sodium intake is 2463 mg. This means the remaining 94.4% of a person's Calorie intake would be limited to have only 1733 mg of Na^+ along with it. If the rest of the day's Calorie intake matched this food in sodium content, the total Na^+ ingested would be 3660 mg Na^+ because $\frac{730 \text{ mg Na}^+}{100 \text{ Calories}} = \frac{3660 \text{ mg Na}^+}{1800 \text{ Calories}}$. This would exceed to recommended 2463 mg by almost 1200 mg Na^+. This amount of salt, Na^+, is not consistent with a low salt diet.

Speaking of Chemistry

Nutrition: The Basis of Healthy Living

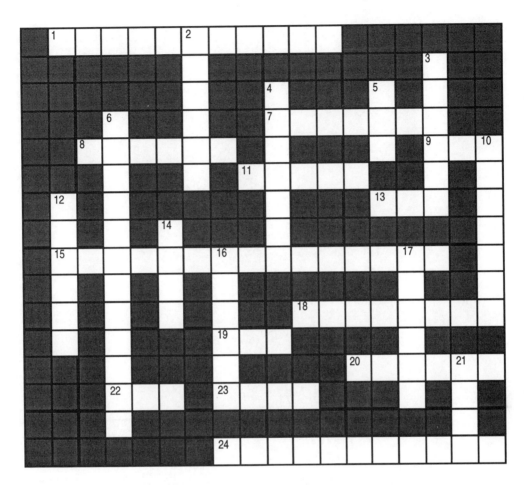

Across
1. Combines with proteins to form lipoproteins
7. Enzyme in saliva that starts breaks down of starch
8. Contains enzyme amylase
9. Monosodium glutamate
11. Indigestible polysaccharides
13. Food and Drug Administration
15. Heart disease
18. Very large vitamin dose
19. Adenosine triphosphate or _____
20. Odor of products from oxidation of unsaturated fatty acids
22. BMR = body weight multiplied by __
23. GRAS list additives are considered
24. Vitamin E prevents oxidation and acts as an _____

Down
2. Deficiency of vitamin C
3. Iron deficiency disease
4. Reactive species with an unpaired electron
5. Yields 9C/gram when metabolized
6. Sugar
10. Body's major fuel molecule
12. Material deposited on artery walls that results in heart disease
14. The generally recognized as safe list.
16. Glucose, fructose, sucrose, etc.
17. Prevents, goiter, enlargement of thyroid
21. Prevents anemia

© 1999 Harcourt Brace & Company. All rights reserved.

Bridging the Gap

Name _____

Food additives, do they help or do they hurt?

The quality and safety of food and medicine has been a concern for people since the beginning of civilization In Western Europe, in 1202 AD, King John of England proclaimed the first English food quality law, the Assize of Bread, which prohibited the adulteration of bread with ingredients like ground peas or beans. Regulations in the United States date back to the colonial period. The State of Massachusetts enacted the first general food adulteration law in the United States in 1848. The need for regulation is accepted. The issue is, "How much regulation?".

This exercise has three parts. You may do this alone or as part of a discussion group. One part is to write a brief argument for continuing to allow food additives. Another part of your assignment is to write an argument for barring food additives. Lastly you are to write two questions you would want answered by expert advisors before you would make a final decision about these two positions.

1. Argument for allowing food additives

2. Argument for barring food additives.

3. Your questions regarding the two different actions.

13 Chemistry and Medicine

13.1 Medicines, Prescription Drugs, and Diseases--The Top Tens

Over-the-counter and prescription drugs are defined. The top ten prescription drugs in the first quarter of 1996 are tabulated. The meanings of the trade name (brand name) and generic name for prescription drugs are explained. The change in the types of drugs prescribed and the maladies treated is traced from 1900 to 1993. The leading causes of death in 1995 are tabulated in Table 13.2

Objectives
After studying this section a student should be able to:
- give the definition for over-the-counter drug
- give the definition for prescription drug
- name the condition for which most prescriptions are written
- identify the classes of drugs that are most prescribed

Key terms
- over-the-counter drugs
- prescription drug
- generic name
- systematic name
- U.S. Food and Drug Administration
- trade name

13.2 Drugs for Infectious Diseases

The history of chemotherapy dates back to Paul Erlich in 1904. Antibiotics are defined as substances produced by microorganisms that inhibit the growth of other microorganisms. A brief history tells of the discovery penicillin G in 1928 and its first clinical use in 1940. The structures and modes of action for penicillins, tetracylines and cephalosporins are described. Guidelines are outlined for minimizing the growth of drug resistant pathogens

Objectives
After studying this section a student should be able to:
- give a definition for chemotherapy, antibiotic, penicillins, cephalosporins, tetracylines
- explain what is meant by a prodrug
- explain how all penicillins function to kill growing bacteria
- tell how cephalosporins and tetracyclines control bacteria
- tell what it means when we say a bacteria is "antibiotic resistant"
- explain why "resistant" bacteria are a public health problem

Key terms

chemotherapy	antibiotics	prodrugs
penicillins	antimicrobial	pathogen
tetracyclines	penicillin G	cell wall development
rifampin	cephalosporin	protein synthesis
RNA synthesis	antibiotic resistance	

13.3 AIDS, A Viral Disease

The history and scope of the spread of acquired immune deficiency syndrome (AIDS, caused by the virus HIV) is discussed. The structure of the HIV retrovirus is described. The steps in the process of HIV invasion of a T lymphocyte are illustrated. A model for the action of AZT in retarding the progression of AIDS is described.

Objectives

After studying this section a student should be able to:
- give the definition for HIV
- explain what is meant by a retrovirus
- describe the process in which an HIV virus attacks a T lymphocyte and causes the production of a new virus
- tell how AZT alters the HIV synthesis of its DNA
- name the base that AZT is believed to replace in DNA
- explain why AZT is not a cure for AIDS
- identify two problems that investigators face when doing AIDS research

Key terms

acquired immune deficiency syndrome	AIDS
Human immunodeficiency virus, HIV	retrovirus
RNA-directed synthesis of DNA	DNA
DNA-directed synthesis of RNA	reverse transcriptase
T cells	AZT, azidothymidine

13.4 Steroid Hormones

Hormones and their interaction with receptors are defined. The structures for natural and synthetic female sex hormones are illustrated. The most commonly used hormones in oral contraceptives are identified as estradiol and ethynyl estradiol. The reasons for prescribing Premarin for women after menopause are explained.

Objectives

After studying this section a student should be able to:
- give the definition for hormone
- give the definition for a receptor
- sketch the shape of the basic steroid carbon skeleton
- tell what development made the pill possible
- name the two most common synthetic hormones used in oral contraceptives
- explain why Premarin is prescribed
- distinguish between the structures for estradiol and ethynyl estradiol

Key terms
- steroid hormones
- chemical messengers
- progesterone-like
- hormones
- oral contraceptives
- Premarin
- receptor
- estrogen-like

13.5 Neurotransmitters

Neurotransmitters are defined. Regulating functions of serotonin and norepinephrine are described. Physiological effects of the combination of both compounds are given. Structures for both compounds are illustrated. The roles of both compounds in the function of nerve synapses are summarized. The common monoamine structural group, $-NH_2$, is identified in both neurotransmittors and antidepressants. The blood-brain barrier is illustrated in Figure 13.4 and its protective function is explained. Parkinson's disease and its treatment with the prodrug L-dopa are described. The link between excess dopamine and schizophrenia is given. Three types of antidepressant drugs are described in terms of their function: tricyclic antidepressants (Elavil), monoamine oxidase inhibitors, those that prevent serotonin recapture by the original neuron, (Prozac). These drugs are identified as SSRI drugs. The use of epinephrine to treat acute asthma attacks and anaphylactic shock from acute allergy attacks is discussed.

Objectives
After studying this section a student should be able to:
- tell what body functions norepinephrine controls
- tell what body functions serotonin regulates
- tell what activity serotonin and norepinephrine appear to control together
- summarize the steps in the cycle of impulse transmission in a nerve synapse
- tell what structural feature all neurotransmitters have in common
- describe the benefits provided by the blood-brain barrier
- tell why L-dopa instead of dopamine is used to treat Parkinson's disease
- describe the physiological effects that epinephrine produces
- name the conditions that epinephrine is used to treat

Key terms
- neurotransmitters
- monoamines
- monoamine oxidase inhibitors
- epinephrine
- Parkinson's disease
- serotonin concentration
- serotonin
- amino group
- Prozac
- dopamine
- schizophrenia
- allergic reactions
- norepinephrine
- monoamine oxidase
- adrenalin
- L-dopa
- tricyclic antidepressants
- anaphylactic shock

13.6 The Dose Makes the Poison

Doses are commonly described in units of milligrams per kilogram of body weight. The dose that kills 50% or half of a large population of test subjects is defined as the LD_{50}. Frequently LD_{50} values are given for rats and extrapolated (estimated) for humans. Difficulties in establishing LD_{50} values are discussed. Species differences in LD_{50} for the same compound are explained. Three routes of entry to the body are identified: inhalation, ingestion, and skin contact. Table 13.3 tabulates LD_{50} values for a variety of compounds from aspirin to lead and nicotine.

Objectives
After studying this section a student should be able to:
- give definitions for dose, lethal dose, sublethal dose, LD_{50}
- explain how human lethal doses can be established without human testing
- explain why it is difficult to estimate risk to humans using LD_{50} values from animal tests
- calculate the total dose that would pose a 50% chance of killing a subject given body mass and LD_{50}
- tell which compound in a set poses the greatest risk if given the LD_{50} values for each
- name the six classifications of toxic substances

Key terms
dose	poison	lethal dose
LD_{50}	species differences	animal data
milligram / kilogram	mg/kg	modes of entry
inhalation	ingestion	skin contact

13.7 Painkillers of All Kinds

Analgesics are defined as pain killers. Three classes of analgesics (over-the-counter-, prescription-, and illegal) are described. Alkaloids are defined as organic compounds that contain nitrogen, are bases, and are produced by plants. The opium poppy is identified as the main source of opium which is a mixture of at least 20 different alkaloids. Morphine is named as the most abundant alkaloid in opium. Opoids are defined as compounds with morphine like properties. Mild analagesics are described. Aspirin's structure synthesis and properties are discussed Structures and properties of other over-the-counter antipyretic and anti-inflammatory agents such as acetaminophen (Tylenol) and Ibuprofen are given.

Objectives
After studying this section a student should be able to:
- state the definition for alkaloids
- tell what compound is primarily responsible for the effects of opium
- name two compounds derived from morphine
- give the definitions for analgesic, antipyretic, and anti-inflammatory
- explain why aspirin can cause stomach bleeding
- tell why aspirin tablets develop a vinegar-like odor
- name three nonaspirin pain relievers
- tell which analgesics contain a carboxylic acid group

Key terms
analgesics	alkaloids	aspirin
heroin, diacetate ester of morphine	morphine	anti-inflammatory
codeine, methyl ether of morphine	opium	opoid
mild analgesics	Demerol, meperidine	naproxen, Aleve
acetaminophen, Tylenol	buffered aspirin	antipyretic
Ibuprofen, Advil and Nuprin		

13.8 Mood-Altering Drugs, Legal and Not

The effect of depressants on the nervous system is described. Structures and the mode of action for barbiturates, phenobarbital, and benzodiazeprines (Valium) are described. Stimulants such as amphetamines, cocaine, and crack are illustrated and described. Natural sources of hallucinogens like LSD, marijuana and mescaline are discussed. The structure and effects of PCP, "angel dust", are described. The DEA drug classification system is explained and listed in Table 13.2.

Cocaine

Objectives

After studying this section a student should be able to:
- state the definitions for stimulants, depressants, and hallucinogens
- give an example of each of the following: a stimulant, a depressant, an hallucinogen
- tell what is meant by a scheduled substance
- name a schedule 1 substance and name its applications
- name a schedule 2 substance and name its applications
- name a schedule 3 substance and name its applications
- tell what GABA does in the nerve synapse
- explain how crack is prepared and tell why it is more addictive than cocaine
- describe how depressants like barbiturates act to inhibit nerve firing and prolong the binding of GABA to its receptors
- explain why a combination of alcohol and barbiturates is more dangerous than each is separately
- name the natural sources of lysergic acid diethylamide, LSD
- name the natural source of marijuana and identify the active agent in marijuana
- tell what amphetamines are, why they are restricted, and identify previous applications
- describe the physiological effects of the following: THC, alcohol, LSD, PCP, crack

Key terms

USDEA	mood-altering drugs	depressants
Schedule 1 drugs	Schedule 2 drugs	insomnia
gamma-aminobutyric acid, GABA	sedation	benzdiazepines
angel dust, PCP, phenylcyclidine	methamphetamine	anxiety
tranquilizers	barbiturates	ethanol
mescaline	Librium and Valium	amphetamines
lysergic acid diethylamide, LSD	stimulants	caffeine
tetrahydrocannabinol, THC	cocaine	narcolepsy
crack	marijuana	hallucinogen

13.9 Colds, Allergies, and Other "Over-the-Counter" Conditions

The connection between allergic reactions, hay fever, and histamines is described. The structures of histamine and antihistamines are given. Decongestants, antitussives, and expectorants are defined. A summary of recommended guidelines for selecting OTC products completes the section.

Objectives
After studying this section a student should be able to:
- give definitions for histamines, antihistamines, decongestants, expectorants
- tell why it is a good idea to know the physiological effects an over the counter drug
- give an example for each of the following:
  - decongestants, antitussives, expectorants, analgesics, antihistamines
- state the four recommended guidelines for selecting OTC products

Key terms

OTC, over-the-counter	decongestants	antihistamines
histamine	allergic symptoms	antitussives
expectorants		

13.10 Preventive Maintenance: Sunscreen and Toothpaste

This section reviews the nature of ultraviolet light. It describes how the skin responds to ultraviolet light to produce melanin. The relation between stratospheric ozone and UV levels is explained. The UV index developed by the National Oceanic and Atmospheric Administration (NOAA) is described. These UV index values are included with the weather reports in 58 cities. The purpose of sunscreens and their mode of action is explained. The sun protection factor (SPF) is defined. The role of the FDA in regulating SPF products is described. The cancer risk posed by tanning and ultraviolet exposure is explained. Formulations of toothpaste are described. The two major compounds in tooth enamel are identified as calcium carbonate ($CaCO_3$) and calcium hydroxy phosphate ($Ca_{10}(PO_4)_6(OH)_2$). The origins of plaque and tartar are examined. Toothpaste contains $Na_4P_2O_7$ to fight plaque, control tartar and fight gum disease. The reason for adding stannous fluoride, SnF_2, to toothpaste is explained.

Objectives
After studying this section a student should be able to:
- tell how the energy carried by ultraviolet light compares with the energy of visible light
- describe how the skin responds to exposure to UV light
- tell how melanin concentrations and skin color relate
- give the definition for erythema
- explain how stratospheric ozone concentrations are related to UV levels at ground level
- tell how SPF values for sunscreens are determined
- explain what sunscreens do to protect the skin from sunlight
- explain what the EPA and NOAA do regarding UV exposure
- tell why a UV index was developed by NOAA
- tell what two main types of substances are present in toothpaste
- tell what role calcium carbonate and calcium hydroxy phosphate play in tooth enamel
- describe what happens when acid contacts tooth enamel
- describe how dental plaque and tartar form and tell how they differ
- explain why sodium pyrophosphate, $Na_4P_2O_7$, is a toothpaste ingredient
- name two abrasives and tell why they are in toothpaste
- explain why a surfactant or detergent is an ingredient in toothpaste
- name two sources of fluoride ion and explain how fluoride hardens tooth enamel
- describe how tooth loss from gum disease and from tooth decay compare

Key terms

ultraviolet (UV) radiation	wavelength	melanin
PABA	nm, nanometer	sun protection factor, SPF
erythema, skin irritation	absorption of light	cumulative effects
stratospheric ozone	sunscreens	tartar

abrasive
tooth enamel
plaque
hydrated silica
stannous fluoride

gum disease
calcium carbonate
calcified plaque
fluoride ion

decay
apatite
$SiO_2 \cdot nH_2O$
fluoride in drinking water

13.11 Heart Disease and Its Treatment

Cardiovascular disease is the number-one killer in America. Heart disease results from atherosclerosis which interferes with blood flow to the heart muscle. The use of diuretics to treat hypertension or high blood pressure is described. Angina, chest pain on exertion, and its treatment with nitroglycerine is described. The purpose and function of cholesterol-lowering drugs, diuretics, vasodilators, beta blockers, tissue plasimogen, and aspirin are explained. Structures for representative compounds like nitroglycerine for angina are given.

Nitroglycerine
$$\begin{array}{l} H_2C-ONO_2 \\ CH-ONO_2 \\ H_2C-ONO_2 \end{array}$$

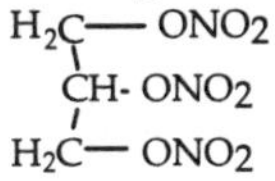

Objectives

After studying this section a student should be able to:
- give the definition for cardiovascular disease
- describe the symptoms for angina
- tell what is meant by the term "heart failure"
- explain why cholesterol-lowering drugs are prescribed and name an example
- state the function of a diuretic and give an example
- name a vasodilator and tell why it is prescribed
- state the purpose of beta blockers
- name a clot dissolving drug used to treat heart attack victims
- describe how regular doses of aspirin are believed to help prevent heart attack

Key terms

cardiovascular disease
high blood pressure
aspirin
cholesterol-lowering drugs
thiazide
NO
propranol, Inderal
plasmin
streptokinase

plaque
hypertension
angina
blood cholesterol
vasodilators
amyl nitrate
beta-1 receptors
clot dissolving drugs
tissue plasminogen activator, TPA

atherosclerosis
diuretics
blood volume
nitroglycerine
beta blockers
beta-2 receptors
plasminogen
prostaglandins

13.12 Cancer and Anticancer Drugs

Cancer can be initiated by physical damage, biological damage or chemical damage to DNA. A summary of cancer incidence for 11 different body sites is illustrated graphically in Figure 13.10. Cancer manifests itself in at least three ways. (1) The rate of cell growth is uncontrolled. (2) Cancerous cells spread to other organs and follow no bounds. (3) Most cancer cells lose specialization of function. Carcinogenesis is often a 2 step process, initiation and promotion. Cancers are treated by surgery, irradiation and chemicals that kill cancer cells. Two types of cancer chemotherapy are described. Alkylating agents (like nitrogen mustards) that interfere with DNA replication are described. Antimetabolite drugs (like methotrexate) that interfere with cancer cell metabolism are discussed. The reason for using synergistic combinations of drugs is explained.

© 1999 Harcourt Brace & Company. All rights reserved.

Objectives
After studying this section a student should be able to:
- describe the two stages in carcinogenesis
- list three ways that cancers are treated
- name the two types of cancer chemotherapy drugs
- tell why alkylating agents are used to treat cancer even though they attack normal cells
- explain why combination chemotherapy is now used instead of single drugs

Key terms

cancer	DNA damage	physical
biological	chemical agents	treatment
surgery	irradiation	chemicals
cancer incidence	cancer death rates	mustard gas
nitrogen mustards	alkylating agents	alkyl groups
antimetabolites	initiation	promotion

Suggested Additional Readings and Resources

Aharonowitz, Yair and Gerald Cohen. "The Microbiological Production Of Pharmaceuticals" Scientific American Sept. 1981: 40.

Ashby, J. and H.Tinwell. "Is Thalidomide Mutagenic?" Nature June 8, 1995: 453.

Cohen, Jon. "Bringing AZT to Poor Countries" Science Aug. 4, 1995: 624.

Gutfield, Greg. "Penicillin's No Villain: You May Not be Allergic to it After All" Prevention Jan. 1992: 18.

Lawn, Richard M. "Lipoprotein(a) In Heart Disease" Scientific American June 1992: 54.

Lipkin, Richard. "Tamoxifen Puts Cancer on Starvation Diet" Science News Nov. 5, 1994: 292. 182.

Liotta, Lance A. "Cancer Cell Invasion and Metastasis" Scientific American Feb. 1992: 54.

Melzack, Ronald. "The Tragedy Of Needless Pain" Scientific American Feb. 1990: 27.

Musto, David F. "Opium, Cocaine And Marijuana In American History" Scientific American July 1991: 40.

Rozin, Skip. "Steroids: a Spreading Peril" (includes related article on steroid use by Chinese swimmers) Business Week June 19, 1995: 138.

Tarpy, Cliff. and José Azel. "Straight: A Gloves-off Treatment Program" National Geographic Jan. 1989: 48.

White, Peter T. and José Azel. "Coca--An Ancient Herb Turns Deadly" National Geographic Jan. 1989: 3.

Winter, Ruth. "Which Pain Relievers Work Best?" (includes related information on drug interactions) Consumers Digest Sept.-Oct. 1994: 76.

Answers to Odd Numbered Questions for Review and Thought

1. Malaria, pneumonia, bone infections, gonorrhea, gangrene, tuberculosis, and typhoid fever are bacterial diseases. Polio, AIDS, and Rubella (German measles) are viral diseases.

3. The United States Food and Drug Administration, FDA, has the responsibility for classifying drugs as either over-the-counter drugs or prescription drugs.

5. Three major classes of antibiotics are penicillins, cephalosporins, and tetracyclines.

   Penicillin G

   Cefaclor, a cephalosporin

   Tetracycline

7. Chemotherapy is the treatment of disease with chemical agents.

8. A retrovirus is a virus that uses RNA directed synthesis of DNA instead of the usual DNA-directed synthesis of RNA.

   [9.]

11. a. Vasodilators are used to treat heart disease and asthma. The vasodilator relaxes the walls of blood vessels and creates a wider passage for blood flow. This lowers blood pressure and reduces the amount of work that the heart must do to pump blood. It also enables a person to breathe more deeply and easily by dilating the bronchial tubes.
    b. Alkylating agents are used to treat cancer. Alkylating agents react with the nitrogen bases of DNA in cancer cells and normal cells. Alkyl groups are added to the nitrogens in the bases. This has an effect on both cancer cells and normal cells, but cancer cells are usually dividing and duplicating DNA more often, so the cancer cells are impacted more.
    c. Beta blockers are used to treat heart disease. Propranolol (Inderol) is used to treat angina, cardiac arrhythmia, and hypertension. Beta blockers act to keep epinephrine and norepinephrine from stimulating the heart.

       Inderal, Propranolol

    d. Antimetabolites are used to treat cancer. Antimetabolites interfere with DNA synthesis, and cancer cells are more susceptible than normal cells because the cancer cells are generally replicating DNA more frequently than normal cells.

13. Histamines and neurotransmitters bind to receptor sites.

15. a. Lovastatin is a cholesterol lowering drug.
    b. Methotrexate is an antimetabolite used to treat cancer.
    c. Chlorphenirimine (in Chlortrimeton) is an antihistamine
    d. Pseudoephedrine is a decongestant.

© 1999 Harcourt Brace & Company. All rights reserved.

17. Codeine, $C_{18}H_{21}NO_3$,
    a. is an analgesic; It is used to relieve pain.
    b. is not an antibiotic; It has no ability or use in limiting pathogen growth.
    c. is an opoid; It has morphine-like activity.

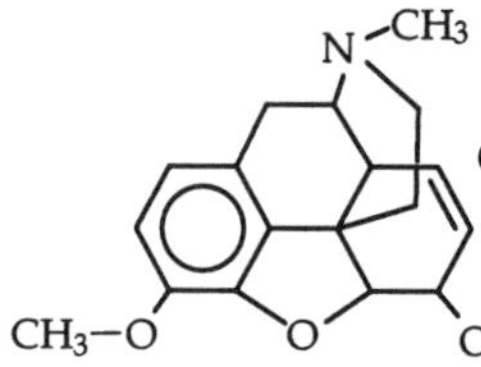

    Codeine

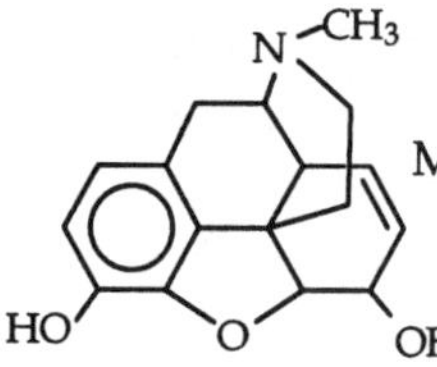

    Morphine

    d. is a controlled substance; It is specifically mentioned as a Schedule 2 drug.
    e. is not an antitussive.

19. a. Angina results from heart disease with the symptom of chest pain on exertion.
    b. Arrhythmia is a heart disease; its symptom is an abnormal heart rhythm.

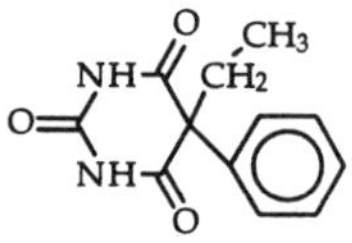

21. A barbiturate such as phenobarbital,

    phenobarbital

    is a depressant. Barbiturates bind to a receptor for GABA. This keeps channels for chloride ion, Cl^{1-}, transmission open. This inhibits transmission of nerve impulses. The physiological effects are a progression from sedation or relaxation, to sleep, to general anesthesia, to coma and death.

23. Heart muscle contracts more often and heart rate increases.

25. Estradiol has an alcohol group and a hydrogen on a carbon in the 5 membered ring while ethynyl estradiol has both an alcohol group and an ethynyl group on that carbon.

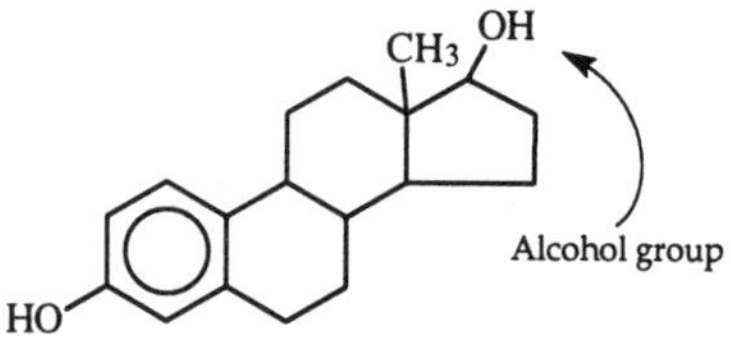

    Estradiol

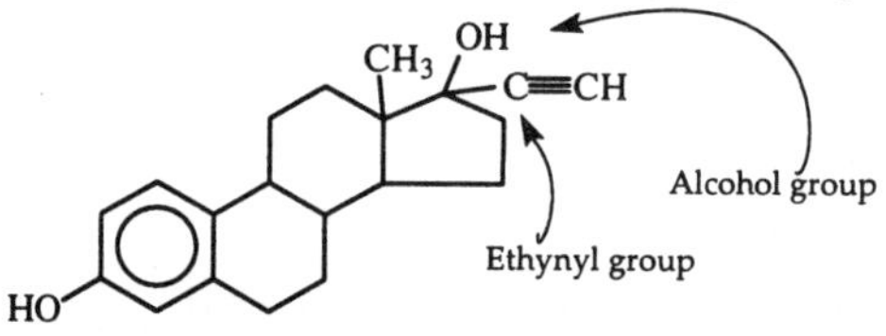

    Ethynyl estradiol

27. a. Dopamine helps control memory, emotion and regulate fine muscle movement.
    b. Epinephrine is a neurotransmitter that produces increased blood pressure, dilation of blood vessels and widening of the pupils of the eye.

29. The four Drug Enforcement Administration drug classifications are: over-the-counter, prescription, unregulated nonmedical drugs, controlled substances.

31. a. Mescaline is a hallucinogen.
    b. LSD, lysergic acid, diethylamide is a hallucinogen
    c. *Cannabis sativa* is a hallucinogen
    d. Phencyclidine (PCP) is a hallucinogen
    e. Barbiturates are depressants
    f. Amphetamines are antidepressants or stimulants

33. a. heart disease — Drugs like cholestyramine and lovastatin act to reduce cholesterol blood levels. Cholestyramine accelerates cholesterol excretion while lovastatin interferes with cholesterol synthesis in the liver.
    b. angina — Vasodilators work to expand or dilate veins. This decreases resistance to blood flow and the blood pressure goes down. This reduces the work the heart must do to circulate the blood.
    c. hypertension — Diuretics reduce blood pressure and hypertension by stimulating excretion of Na^{1+} and increasing urine production. Blood volume and blood pressure decrease when urine output increases.

35. Surgery, irradiation, chemotherapy

37. Hallucinogens cause a person to experience vivid illusions, fantasies, and hallucinations. Examples of hallucinogens are mescaline, PCP, and LSD.

39. a. Morphine is a more effective pain killer than heroin and codeine.
    b. Heroin is not a natural alkaloid.
    c. Heroin is so addictive that it is not legal to sell or use it in the United States.

41. Nitroglycerine is a heart muscle relaxant. It is used to treat angina which is a symptom of heart disease.

    Nitroglycerine:
    $$\begin{array}{l} H_2C-ONO_2 \\ CH-ONO_2 \\ H_2C-ONO_2 \end{array}$$

43. The blood-brain boundary is defined physically by small openings in capillaries and the astrocyte cell membranes that prevent passage of large polar molecules. The barrier keeps large toxic polar molecules from passing into the brain. Drugs must be soluble in blood and soluble in the lipid layer of the membrane in order to reach the brain.

Speaking of Chemistry

Chemistry and Medicine

Name_____

Across
4 Formed from L-dopa
7 Substances with morphine-like properties
8 Treated using antipyretics
9 Chest pain caused by exercise
10 Methotrexate similar to folic acid, interferes with DNA synthesis
13 One of first FDA approved drugs produced using recombinant DNA
15 Methyl ether of morphine
17 Gamma-aminobutyric acid
20 Antibiotic isolated in 1940, discovered by Sir Alexander Fleming in 1928.
22 Caused by carcinogens

Down
1 Food and Drug Administration
2 Infectious disease caused by HIV
3 Organisms that are disease causing
5 Drugs that are misused
6 Derived from coca plant leaves
8 Methotrexate, interferes with reduction of _____ acid
10 Zidovudine used with a protease inhibitor to treat AIDS
11 Once was extracted from willow bark
12 Over the counter drugs
14 Acts to dissolve blood clots
16 Treated using analgesics
18 Sun protection factor
19 Angel dust
21 Lysergic acid diethylamide

© 1999 Harcourt Brace & Company. All rights reserved.

Bridging the Gap   I   Name _____
Internet access to the Food and Drug Administration, FDA
FDA Cancer Drug Evaluation and Approval

FDA Cancer Drug Marketing Approval

The United States Food and Drug Administration has the responsibility for reviewing and approving drugs before they can be sold in the USA. The FDA must decide on the safety and effectiveness of each new drug.

This exercise is intended to familiarize you with the Food and Drug Administration, FDA. The excercise deals with the time required to bring new cancer drugs to market and the number of drugs approved in 1994. This exercise gives you information about the FDA policy changes and programs to increase availability of cancer drugs to patients in the United States. You should be able to answer the following questions after viewing this FDA site.

    http://www.fda.gov/opacom/factsheets/cancerfs.html

1. What does the abbreviation FDA mean?

2. What is the average number of months the FDA requires to review and approve cancer drugs? Does the wording of the site imply this is typical for FDA approval?

3. How many drugs were approved for cancer treatment in 1994?

   What cancer drugs were approved for patient use in 1994?

4. Each approved drug is specific for one type of cancer. What cancer conditions do these drugs treat? Why do you think different drugs are needed for each type of cancer?

Another site that deals with cancer treatment and chemistry has the following URL. The effectiveness of DMSO and Polar/Planar molecules in treating malignancy is addressed at this site.
    http://oncolink.upenn.edu/specialty/chemo/drugs/dmso_1.html

© 1999 Harcourt Brace & Company. All rights reserved.

Bridging the Gap II Name _____
FDA Prescription Drug Evaluation and Drug Prices

This is a two part exercise. The first part deals with the time required to bring a new drug to market. The second deals with prescription drug prices.

1. FDA Prescription Drug Evaluation: Too Slow or Just Right

The United States Food and Drug Administration has the responsibility for reviewing and approving drugs before they can be sold in the USA. The FDA must decide on the safety and effectiveness of each new drug. This process can take as long as 12 years from the initial discovery to final approval and public sale. Many countries like France and Great Britain have equally thorough drug evaluation procedures that are much quicker.

Very often a potential drug is discovered by a U.S. pharmaceutical company. The company starts the evaluation process simultaneously in the United States and overseas. The testing process will be completed overseas and the drug is approved for sale while the FDA is still doing its review. The drug is typically unchanged when the FDA approves it for use in the USA a few years later. In the meantime the drug is not available to the public in the USA.

Your assignment is to discuss this situation with your classmates, family or friends. You are supposed to develop two questions you would want answered about the difference in evaluation times. Write your two questions in the space below.

1.

2.

© 1999 Harcourt Brace & Company. All rights reserved.

2. Prescription Drug Prices

Prescription drugs can be very expensive. There are trade name prescription drugs and generic name prescription drugs. The two are typically very different in price; the generics usually cost less.

Your assignment has two parts. The first part is to create a question you would ask a pharmacist or health care professional about the differences between generic and trade name prescription drugs. Then you should talk to a health care professional and actually ask your question. Record both your question and a summary of your conversation with the health care professional. The second part is to discuss this situation with your study group, classmates, family or friends. Based on these discussions develop an opinion about why the generics cost less than the trade names.

1. Question for health care professional

2. Summary of conversation with health care professional

3. Summary of your discussions and your opinion

14 The Chemistry of Useful Materials

14.1 The Whole Earth
This section describes the structure and composition of the earth's crust. Definitions are given for the three major types of rocks: igneous, sedimentary, and metamorphic. Minerals are defined and the forces and processes that distribute minerals in the crust are described. Regions of the world that have high concentrations of specific minerals are named. Mineral groups found in the earth's crust are tabulated. A cross sectional diagram illustrates the relationships between the earth's lithosphere, asthenosphere, mantle and core.

Objectives
After studying this section a student should be able to:
- give a definition for the hydrosphere
- give definitions for igneous, sedimentary, and metamorphic rock
- state the definition for minerals
- give an approximate value in kilometers for the thickness of the earth's crust
- name two major mineral groups, give a specific example of a mineral in the group

Key terms
igneous rocks	hydrosphere	Earth's crust
silicates	sedimentary rocks	metamorphic rocks
major mineral groups	minerals	lithosphere

14.2 Chemicals from the Hydrosphere
This section describes how useful materials can be extracted from the hydrosphere. The chloralkali process to produce chlorine and sodium hydroxide from brines is described. The production of magnesium metal from sea water is diagrammed.

Objectives
After studying this section a student should be able to:
- define brines and salts
- tell what the chloralkali industry produces and describe the chloralkali process
- state what raw materials the chloralkali industry uses
- explain why mercury cathodes in the chloralkali process pose an environmental hazard
- define the term alloy and give an example
- explain how precipitation is used to separate magnesium from sea water
- tell how $CaCO_3$ from seashells is used to make lime, CaO

Key terms
brines	precipitation	chloralkali industry
calcium carbonate	electrochemical cell	electrolysis of aqueous NaCl
alloy	lime	insoluble

14.3 Metals and Their Ores
This section describes common the connection between minerals, ores and refined metals. Iron production and the function of the blast furnace to reduce iron oxides is discussed.. Compounds and reactions in the reduction process are identified. Steel production is described. The history of aluminum refining is recounted. Reactions in the copper extraction process are discussed and reactions for copper production are explained.

Objectives

© 1999 Harcourt Brace & Company. All rights reserved.

After studying this section a student should be able to:
- state a definitions for free elements, slag, alloy, steel, ores, pig iron, cast iron, bauxite
- tell what a blast furnace does to a mixture of iron oxide (Fe_2O_3), coke (C), and limestone ($CaCO_3$)
- tell what pig iron is
- tell how cast iron differs from pig iron
- describe how steel differs from cast iron

Key terms

free elements	reactive metals	sulfides
oxides	chemical reduction	iron oxide
blast furnace	silica	slag
pig iron	cast iron	cementite
steels	carbon steel	basic oxygen process
refractory	electrolysis	corrosion resistant
flotation process	ductile	

14.4 Conductors, Semiconductors, and Superconductors

This section describes the characteristics of metals, semimetals (semiconductors) and superconductors. Models that explain these properties are introduced. Examples are used to explain how a p-type and n-type semiconducter are made. Applications for semiconductors and superconductors are discussed.

Objectives

After studying this section a student should be able to:
- describe electrical conductivity, thermal conductivity, ductility, and luster of metals
- tell what the "electron sea" is in metals
- give an example of a semiconductor and tell how they differ from metals
- tell what doping of nonmetal crystals does to the crystal structure and conductivity
- give definitions for p-type and n-type semiconductors
- describe the combination of semiconductors used to assemble a transistor
- define a superconductor
- tell how MRI machines and superconductors are related

Key terms

ductility	electrical conductivity	thermal conductivity
insolubility	malleability	luster
valence electrons	lattices	sea of mobile electrons
doping	semiconductors	electron gate
extra electrons	dopant	magnetic resonance imaging
transistor	p-type semiconductor	n-type semiconductor
superconductor	positive holes	superconducting transition temperature
MRI		

14.5 From Rocks to Glass, Ceramics, and Cement

This section describes the structure for silica, SiO_2, and silicates. The structures for the chains of linked tetrahedra in silicates (pyroxenes) and side by side linked chains in amphibole (asbestos) are illustrated. The types of asbestos and the hazards they pose are discussed in detail. Glass materials are hard noncrystalline transparent substances with an internal structure like a liquid. There are no regular repeating structural units. This model accounts for the irregular fractures observed for glasses. Annealing is a process of slow cooling that minimizes stress and equalizes bonding forces in the glass. Amorphous materials have no regular structure. The reactions between Na_2CO_3 and SiO_2 to produce glass and the glass manufacturing process are described. Ceramics are defined, the properties and applications of ceramics are described. Definitions are given for cement and concrete. The production and composition of Portland cement are described.

Objectives

After studying this section a student should be able to:
- give a definition for silicates, silica, clays, glass
- sketch a tetrahedron
- name the two types of asbestos and tell which is more hazardous
- give a definition for amorphous substances
- describe the purpose and reason for annealing glass
- state a definition for ceramics and tell why they are economically attractive
- describe composite materials
- give a definition for cement
- tell how mortar differs from concrete
- give a definition for Portland cement

Key terms

silica	pyroxenes	tetrahedra
amphibole	fibrous natural silicates	crocidolite
asbestos	chrysotile	EPA
glass	amorphous	annealing
Pyroceram	ceramics	ceramic composites
cement	Portland cement	concrete

Additional Readings

Amato, Ivan. "New Superconductors: a Slow Dawn" <u>Science</u> Jan. 15, 1993: 306

Bowen, H. Kent. "Advanced Ceramics" <u>Scientific American</u> Oct. 1986: 168.

Coy, Peter. "Man-made Diamonds Learn a New Trick from Mother Nature" <u>Business Week</u> June 7, 1993: 103.

Keyes, Robert W. "The Future of the Transistor" <u>Scientific American</u> June 1993: 70.

Pool, Robert. "Atom Smith. (Dick Siegel Constructs Materials One Molecule at a Time)" <u>Discover</u> Dec. 1995: 54.

Wallach, Jeff. "Self-healing Concrete" <u>Popular Science</u> Feb. 1993: 19.

Answers to Odd Numbered Questions for Review and Thought

1.
   a. The hydrosphere is the layer of fresh water and salt water above and below the Earth's surface
   b. Igneous rock is rock formed by solidification of a *magma* or molten rock
   c. rock formed by deposition of dissolved or suspended substances

3.
   a. Annealing is the process of heating and then slowly cooling a substance to make the substance less brittle and reduce strain in the solid.
   b. Amorphous solids have no regular crystalline order.
   c. Ceramics are materials generally made from clays and then hardened by heat.
   d. Cement is a substance able to bond mineral fragments into a solid mass.
   e. A glass is a hard, noncrystalline substance with random (liquid-like) structure.

5. Magnesium is extracted from sea water. It is used in fireworks, flashbulbs, and alloys for auto and aircraft parts.

7. Gold, copper, and platinum all can be found in the pure metallic element in nature.

9. The slag is lower density than the molten iron. The low density molten slag floats on the higher density molten iron. The two layers are not soluble in each other so they do not mix.

11. Excess carbon is used when Fe_2O_3 is reduced to iron, Fe. This excess carbon remains in pig iron as an impurity. Pig iron is brittle partly because it contains Fe_3C.

13.
    a. $2\ Cu_2S_{(s)} + 3\ O_{2(g)} \longrightarrow 2\ Cu_2O_{(s)} + 2\ SO_{2(g)}$
    b. $2\ FeS_{(s)} + 3\ O_{2(g)} \longrightarrow 2\ FeO_{(s)} + 2\ SO_{2(g)}$

15. Metals are malleable, ductile, good conductors of electricity and heat, lustrous, insoluble in water or other solvents.

Property	Use
ductile	copper wire, aluminum wire, steel rod and cable
malleable	copper sheet, aluminum foil
electrical conductor	gold contacts, copper wire, thermostats
heat conductor	aluminum, copper and steel pots and pans
lusterous	gold and silver jewelry, silver mirrors, chrome plating

17.
    a. The p-type semiconductors have positive holes where there is a shortage of electrons. The p-type semiconductors are made by doping silicon crystal with an element like gallium or boron from Group IIIA. The silicon atoms have four valence electrons and the dopant atoms have only three electrons. This creates a shortage of electrons in the crystal.
    b. The n-type semiconductors have excess electrons and negative holes. Silicon, with four valence electrons, is doped with elements like arsenic and phosphorus from Group VA. These have five valence electrons so the doped crystal has extra electrons.

19. The structure of an SiO_4 unit is tetrahedral.

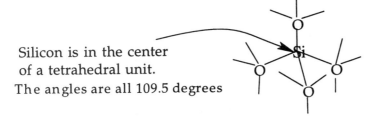

Silicon is in the center of a tetrahedral unit.
The angles are all 109.5 degrees

21. A superconductor offers no resistance to electron flow below a certain critical temperature called the superconducting transition temperature. Superconductivity is the behavior displayed by a superconductor at temperatures at or below the superconducting transition temperature. Superconductivity research is aimed at minimizing energy losses when transferring electricity at higher and higher temperatures. Present day superconducting transition temperatures are in the 70 to 100 Kelvin range. This is expensive to reach and maintain on a large scale because of the cost of cryogenic liquids (liquid nitrogen) and handling equipment.

23. Iron (III) compounds give glass a yellow color. This can be seen more easily if you look along the edge of a glass plate so you can look through a greater thickness of glass. The color may not be easy to see if you look through a thin layer of glass.

25.

Property	Metal	Ceramic
Hardness	variable	yes
Strength	yes	yes
Ductility	yes	no
Electrical Conductivity	yes	no
Brittleness	no	yes

27. Lime, CaO, reacts with CO_2 to form solid $CaCO_3$. This is an ionic solid.

29. No, when a high temperature mixture of silicates and metallic elements cools, the two parts of the mixture will not all solidify or crystallize at the same time. The silicates will crystallize first because of their higher melting points. The metallic elements will crystallize later at a lower temperature. This leads to separate pockets or veins of silicate and metallic matter.

31. Yes, if the elements were uniformly distributed on the earth it would be more difficult to extract or separate them because we would need to process more material to accumulate the same amount of an element. The separation of useful pure elements would require processing more material and lead to greater disruption of the environment.

Speaking of Chemistry

The Chemistry of Useful Materials

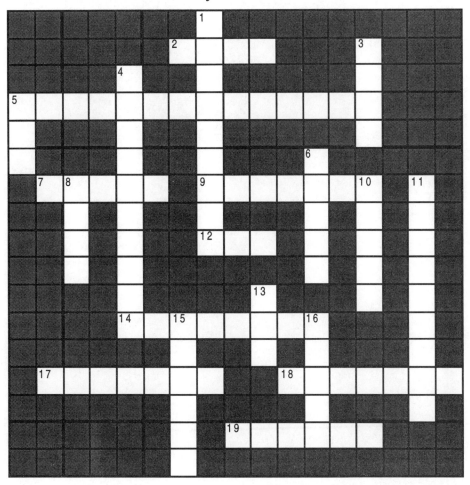

Across
2 Objects formed in molds from pig iron
5 Zero resistance material
7 Solid appearing substance that has no crystalline form
9 Rocks formed from molten rock
12 Magnetic resonance imaging
14 Process of heating an ore to high temperature
17 Property of being able to be drawn into wires
18 Substance that has gained an electron or lost oxygen
19 Salty water

Down
1 Metal element recovered from sea water
3 Resistance of superconductor
4 Electron gate formed using sandwich of p-type silicon between two n-type silicon
5 Source of magnesium salts
6 Places in semiconductor crystal that are electron deficient
8 Calcium oxide, CaO
10 Alloy of iron recovered from basic oxygen process
11 Chains of SiO_2 tetrahedra
13 Form of iron recovered from a blast furnace
15 Homogeneous mixtures of metals
16 Color imparted to glass by traces of iron III, Fe^{3+}, compounds

Bridging the Gap
Tetrahedron Construction and Silicates

The two dimensional illustrations in the chapter are excellent, but they cannot give a tactile sense of the tetrahedral form and the structures for extended silicates. This exercise is intended to give you a hands on experience with a three dimensional model of the tetrahedron and, if you choose, with even larger structures.

Please read all these directions before doing any cutting.

Write your name in the blank space provided on the template. Your instructor may want you to turn in your completed tetrahedron. Be careful to keep the A, B, and D tabs on the template when you cut out the tetrahedron. Be sure to leave the black edges on the faces.

Hold the cutout so you can read your name. Fold faces A, B, and D away from you. Hold face D up so you can read it. Fold the hidden support away from you. Do this same process with face B and the second hidden support. Slide the hidden support behind face A. Fold tab B over the B on face B. Insert the remaining hidden support behind face B. Fold tab A over the A on face A. Fold tab D over the D on face D. All the tabs can be secured with a piece of transparent tape if you wish. You now have your tetrahedron that represents the structure for SiO_4. The oxygens 1,2,3,4 are at the matching corners of the tetrahedron. The silicon atom is in the center of the tetrahedron. The tetrahedron is the shape for the fundamental building block of silicates.

Silicate Anions

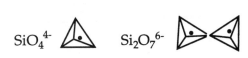

SiO_4^{4-} $Si_2O_7^{6-}$ $Si_3O_9^{6-}$

Silicate Chains and Asbestos

The chain structures for pyroxenes and amphiboles can be built by joining tetrahedra through their corners so an oxygen is shared between two silicon atoms in neighboring units. The chains give a fibrous material. The crocidolite double chain leads to long thin needles that are able to penetrate the airways in the lung. The crocidolite asbestos is less soluble and persists longer in tissue. This increases the risk of lung cancer and other health problems. You and your classmates can duplicate the crocidolite structure by sharing your tetrahedra.

Pyroxenes (linear chains)

Amphiboles (double chains)
This is the structure for crocidolite asbestos.
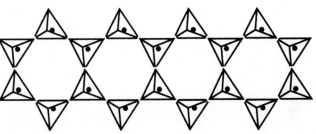

© 1999 Harcourt Brace & Company. All rights reserved.

Bridging the Gap
Template for silicate tetrahedra

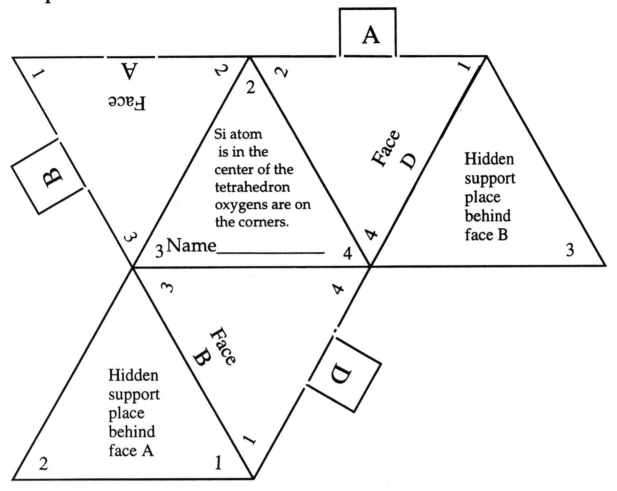

© 1999 Harcourt Brace & Company. All rights reserved.

221

15 Water--Plenty of It, But of What Quality?

15.1 How Can There Be a Shortage of Something as Abundant as Water?
The responsibility of the U. S. Environmental Protection Agency (EPA) to enforce federal laws regulating waste water and drinking water quality is described. Water is identified as the most abundant compound on the earth's surface. The small amount (2.5%) of fresh water and its uneven distribution are discussed. Surface water and groundwater are defined and named as the sources of usable water. Aquifers and artesian wells are described. The Ogallala aquifer in the region between South Dakota and Texas is discussed. Salt water intrusion into coastal fresh water supplies to produce "brackish" water is explained. Ground subsidence and sinkhole formation when aquifers are tapped by deep wells are discussed.

Objectives
After studying this section a student should be able to:
- give definitions for surface water, groundwater, aquifers, artesian wells
- identify the sources of surface water
- describe the problem that deep wells pose
- give the approximate depth of a "deep" well
- explain what is meant by aquifer depletion
- describe the relationship between aquifer depletion and salt water intrusion
- explain how sinkholes are created by pumping water from aquifers

Key terms
surface water	groundwater	aquifers
artesian wells	shallow well	deep well
aquifer depletion	brackish water	sinkholes

15.2 Water Use and Reuse
The major consumers of fresh water are tabulated. Agriculture is identified as the largest single user of fresh water. Industrial use of cooling water is explained. Potable water is defined as drinkable water. Groundwater and surface water are named as the source of potable water. The average water usage per person per day for bathing, laundering drinking, etc., is tabulated. Average individual drinking water consumption is estimated as 2 gallons per day. Natural recycling of water molecules is discussed. Aquifer depletion is described. The replacement of aquifer water by groundwater recharge is explained.

Objectives
After studying this section a student should be able to:
- name the major users of fresh water in the United States
- describe why water is used as a coolant
- give definitions for thermal pollution, potable water, dual water system
- explain why thermal pollution creates problems for aquatic life
- explain why dual water systems are receiving attention
- describe the purpose of groundwater recharge

© 1999 Harcourt Brace & Company. All rights reserved.

Key terms
- industrial water
- dual water systems
- groundwater recharge
- thermal pollution
- wastewater
- potable
- sewage

15.3 What is the Difference Between Clean Water and Polluted Water?

Pollution is defined. The official U.S. Public Health Service classes of pollutants are identified and tabulated. Pollutants are identified as: heat, radioisotopes, toxic metals, acids, toxic anions, organic chemicals, infectious agents (pathogens), sediment from land erosion, plant nutrients, plant and animal matter. The Clean Water Act of 1977 is described. The discharger's responsibility to ensure the cleanliness of wastewater is explained. The role of the EPA is described.

Objectives

After studying this section a student should be able to:
- give the definition for pollution
- name the eight classes of pollutants identified by the U.S. Public Health Service
- explain how The Clean Water Act of 1977 changed responsibility for keeping water clean
- summarize the responsibilities of the EPA regarding water quality

Key terms
- pollution
- toxic metal
- alkalis
- human activity
- pollutants
- heat
- acids
- pathogens
- natural processes
- wastewater effluent
- radioisotopes
- organic molecules
- "pure" natural water
- Clean Water Act of 1977
- Federal Water Pollution Control Act

15.4 The Impact of Hazardous Industrial Wastes on Water Quality

The problems that result from solid waste disposal in "old" style landfills are explained. The water pollution conditions that led to the Resource Conservation and Recovery Act of 1976 (RCRA) are described. The purpose of the "Superfund" to clean up hazardous waste disposal sites is described. The "cradle-to-grave" responsibility of waste generators is discussed. The RCRA requirement for hazardous waste "manifests" is explained. The concept of "secure" landfills is described. The prohibition against putting hazardous waste in landfills is cited as one reason for the incineration of hazardous waste. Legislation like California's Proposition 65 to prohibit the disposal of hazardous materials in the water supply is discussed.

Objectives

After studying this section a student should be able to:
- describe the problems that "old landfills" pose
- describe the situation that led to the creation of RCRA and the "Superfund"
- describe the purpose of the "Superfund"
- identify the agency that defines hazardous wastes
- explain what the term "cradle-to-grave" means
- tell the purpose of a "manifest" for hazardous wastes
- state the three presently accepted ways to dispose of hazardous waste
- describe some of the security features of a "secure landfill"
- describe the purpose of laws like California's Proposition 65

Key terms

solid wastes	landfills	"responsible parties"
RCRA	Superfund	Resource Conservation and Recovery Act
cradle-to-grave	incineration	hazardous wastes
manifest	secure landfills	California Proposition 65

15.5 Household Wastes that Affect Water Quality

Common household products that become hazardous wastes are identified. The hazardous components of household wastes are classified as acids, caustics, organic solvents, etc. (See table 15.8) Problems faced when disposing of household wastes are explained. Methods for reducing the problem of hazardous household waste are discussed.

Objectives

After studying this section a student should be able to:
- explain why the individual small amounts of household wastes can be a major source of pollution
- explain why individual households have problems with the disposal of hazardous waste
- tell how the amount of oil spilled by the Exxon Valdez compares with the amount of motor oil poured down the drain

Key terms

special disposal	disposal down a drain	trash disposal
household wastes	pesticide waste	hazardous household wastes

15.6 Toxic Elements Often Found in Water

This section discusses heavy metal poisons. Common heavy metal poisons like arsenic (As), lead (Pb), and mercury (Hg) are described. The FDA is identified as responsible for setting standards for arsenic levels. The EPA sets allowed levels for lead in drinking water.

Objectives

After studying this section a student should be able to:
- give a definition for heavy metal poisons
- name three common heavy metal poisons
- name sources for arsenic, lead, and mercury
- describe the symptoms of lead poisoning
- tell which forms of arsenic, lead, and mercury are most toxic
- explain why heavy metals are toxic
- draw the structure for a sulfhydryl group

Key terms

heavy metal poisons	mercury (I) ion	mercury (II) ion
mercury	lead	arsenic
EPA	FDA	sulfhydryl groups, -SH
parts per billion, ppb	parts per million, ppm	

15.7 Measuring Water Pollution

Analytical chemistry is defined. Biological oxygen demand, BOD, is defined. The consumption of oxygen by organic waste is discussed. Concentration measures like parts per million (ppm) are explained. The solubility of oxygen in water is discussed. The decrease in O_2 solubility with temperature is linked to thermal pollution. Gases are less soluble in warm water than they are in cold water. The method for calculating BOD is shown. The connection between BOD, dissolved oxygen, and fish kills is described and illustrated.

Objectives

After studying this section a student should be able to:
- give the definition for analytical chemistry
- state the definition for biological oxygen demand, BOD
- name the types of products formed by oxidation of organic carbon
- describe the relationship between BOD and dissolved oxygen
- tell how BOD relates to survival of fish
- tell how the concentration of dissolved oxygen relates to water temperature

Key terms

analytical chemistry	dissolved oxygen	biochemical oxygen demand, BOD
metabolize	burning	dissolved organic matter
oxygen depletion	parts per million	solubility of oxygen
aerobic bacteria	anaerobic bacteria	

15.8 How Water Is Purified Naturally

The seven different natural water purification processes are identified and briefly described. Biodegradable and nonbiodegradable substances are defined. Branched-chain detergents and linear-chain detergents are used as examples of nonbiodegradable and biodegradable organics. Problems posed by hard chlorinated pesticides are described. The fat solubility of DDT and related compounds is discussed. The concentrating effect of the food chain on fat soluble polychlorinated hydrocarbons is described. The reason for banning DDT in the U.S. is explained.

Objectives

After studying this section a student should be able to:
- name seven different natural water purification processes
- state the definitions for biodegradable and nonbiodegradable
- describe how linear chain and branched chain detergents differ in biodegradability
- give the definition for a persistent insecticide
- tell why chlorinated hydrocarbons like DDT are hazardous
- tell how small concentrations of DDT in fresh water can lead to serious health hazards

Key terms

water cycle	distillation	crystallization
aeration	filtration	sedimentation
settling	oxidation	dilution
biodegradable	nonbiodegradable	natural purification
food chain	chlorinated hydrocarbon	polychlorinated hydrocarbon
fat soluble	detergent	branched chain detergent
DDT	linear-chain alkylbenzenesulfonate detergent	

15.9 Water Purification Processes: Classical and Modern

The old fashioned cess-pool method for sewage treatment is described. The evolution of sewage treatment plants is summarized. The levels of sewage treatment are defined: primary treatment, secondary treatment, tertiary treatment. The types of wastes removed by each type of treatment are identified. Special problem pollutants are named. The reasons for carbon black filtration and denitrifying bacteria are explained.

Objectives

After studying this section a student should be able to:
- describe the water purification processes used in primary wastewater treatment
- describe the water purification processes used in secondary wastewater treatment

describe the water purification processes used in tertiary wastewater treatment
tell what kind of wastewater treatment is required by the 1972 Clean Water Act
explain why filtration through carbon black is used in tertiary wastewater treatment
tell why denitrifying bacteria are used in wastewater treatment

Key terms

water purification	outhouses	cesspools
sedimentation	chlorinator	primary wastewater treatment
chlorination	aerobic	1972 Clean Water Act
anaerobic	sludge	secondary wastewater treatment
toxic metal ions	nitrate ions	tertiary wastewater treatment
ammonium ions	carbon black	denitrifying bacteria
adsorbed	adsorption	

15.10 Softening Hard Water

Hard water is defined. The metal ions Ca^{2+}, Mg^{2+}, Fe^{3+} and Mn^{2+} that make water "hard" are identified. The undesirable properties of hard water are identified. Methods for "softening" hard water are described in detail. The health risks associated with the lime-soda water softening process are explained.

Objectives

After studying this section a student should be able to:
state the definition for hard water
name the ions present in hard water
name three problems associated with hard water
describe how to soften hard water that contains Ca^{2+} and Mg^{2+} ions
describe how to soften hard water that contains Fe^{2+} and Mn^{2+} ions
tell what happens to soap when it dissolves in hard water

Key terms

hard water	hardness	lime-soda
calcium carbonate	calcium bicarbonate	magnesium hydroxide
magnesium bicarbonate	soft water	oxidation with air
aeration	pH	precipitate
acidic		

15.11 Chlorination and Ozone Treatment of Water

The effectiveness of chlorination of water in controlling water borne diseases like typhoid is described. The oxidizing power of chlorine is described as the reason for its disinfecting power. Disinfection byproducts are defined. The potential hazards of disinfection byproducts are explained. Disinfection of water with ozone is described. Bromate ion is identified as one of the ozonation disinfection by-products.

Objectives

After studying this section a student should be able to:
tell why gaseous chlorine and ozone are used to treat water
name the water-borne diseases chlorination is used to control
give the definition for disinfection byproducts and tell why they are a problem
describe the problems posed by ozonation and chlorination

Key terms

chlorination	chlorine	oxidizing
cholera	typhoid	water-borne diseases
paratyphoid	dysentery	disinfection byproducts
giardiasis	carcinogens	ozone
mutagenic	ozonation	bromate ion

15.12 Fresh Water from the Sea

The percentage of salts in ocean water is tabulated. The acceptable limit on dissolved salts is given as less than 500 ppm. Reverse osmosis and solar distillation are identified as water purification methods.

Objectives
After studying this section a student should be able to:
- give the definitions for permeable membrane, semipermeable membrane, osmosis, reverse osmosis, osmotic pressure, distillation, still
- tell why desalination is necessary
- diagram an apparatus for producing fresh water from salt water using reverse osmosis
- sketch a diagram of a simple solar still

Key terms

fresh water	parts per million, ppm	solar distillation
reverse osmosis	permeable	semipermeable membrane
membrane	osmosis	osmotic pressure

15.13 Pure Drinking Water for the Home

The properties of bottled drinking water are described. Spring water is defined. The effectiveness of distillation, reverse osmosis and carbon filtration as purification methods is summarized.

Objectives
After studying this section a student should be able to:
- name three commonly accepted methods for water purification
- explain why it is important to read labels on bottled water
- name the kinds of pollutants that pass through purification by reverse osmosis
- name the kinds of pollutants that pass through purification by distillation
- name the kinds of pollutants that pass through purification by carbon filtration

Key terms

bottled water	"spring" water	home water-treatment
trace pollutants	analysis on the label	

15.14 What About the Future?

Increasing national and global water use are described. The need to recycle wastes and waste water is discussed. Problems with present social attitudes toward waste disposal are described. The long term need for new ways to deal with politically sensitive waste disposal issues is discussed. The prevalent NIMBY (not in my back yard) social attitude is described.

Objectives
After studying this section a student should be able to:
- tell why water recycling is actually an old practice
- give reasons why household hazardous waste disposal will continue to be a problem
- explain why conservation and waste reduction will be increasing important in the future
- tell what problems result from the NIMBY response
- explain why water recycling will be more common in the future

Key terms

heavy metals	pesticides	Clean Water Act of 1977
chlorinated organic	monitoring	pollution-discharge regulations
waste reduction	household wastes	politics of water protection
water conservation	protection	water recycling

Additional Readings

Boyle, Robert H. "The Killing Fields" (toxic drainwater) Sports Illustrated Mar. 22, 1993: 62.

Canby, Thomas Y. "Water Our Most Precious Resource" National Geographic Aug. 1980: 144.

Cantor, Kenneth P. "Water Chlorination, Mutagenicity, and Cancer Epidemiology." The American Journal of Public Health Aug. 1994: 1211.

Colson, Steven D. and Thom H. Dunning Jr. "The Structure of Nature's Solvent: Water" Science July 1, 1994 : p 43.

Cooper, Mary H. "Global Water Shortages: Will the Earth Run Out of Freshwater?" CQ Researcher Dec. 15, 1995: 1113.

Farvolden, Robert. "Water Crisis: Inevitable or Preventable?" Geotimes July 1995: 4.

"Fit to Drink" Consumer Report Jan. 8, 1990: 27.

Karp, Jopnathan. "Water, Water, Everywhere" Far Eastern Economic Review June 1, 1995: 54.

Knight, Charles and Nancy Knight. "Snow Crystals" Scientific American Jan. 1973: 100.

Mohnen, Volker A. "The Challenge Of Acid Rain" Scientific American Aug. 1988: 30.

Penman, H.L. "The Water Cycle" Scientific American Mar. 1970: 98.

Ward, Fred. "South Florida Water: Paying the Price" National Geographic July 1990: 89.

© 1999 Harcourt Brace & Company. All rights reserved.

Answers to Odd Numbered Questions for Review and Thought

1.
   a. Surface water is water available in rivers, lakes and streams.
   b. Groundwater is water beneath the earth's surface.
   c. An aquifer is a layer of water-bearing porous rock.
   d. Brackish water is fresh water contaminated by salty sea water.
   e. Pollution is defined as any condition that causes the natural usefulness of water, air, or soil to be diminished.
   f. Groundwater recharge is the process of recycling treated sewage effluent back into the aquifer.
   g. Potable water is safe, drinkable water.
   h. Dilution is the process of decreasing pollutant concentration bey the addition of water.
   i. Hazardous waste is waste material that has the potential to harm the environment as defined by the EPA. These may be organic slvents, toxic metals, pesticides, acids, etc.

3. Most of the rain water that falls on the United States each day returns to the atmosphere by evaporation or transpiration from plants.

5. Groundwater can be contaminated with pollutants when rainwater runs over or filters through materials, dissolves the pollutants, then percolates into the ground water.

7. Agriculture is the largest single user of water in the United States.

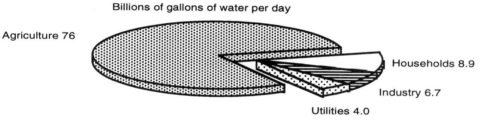

9. Water use per day differs from person to person. Measured in gallons per day an average person is likely to use the following amounts.

Use	Gallons
flushing toilets	30
Bathing	23
Laundering	11
Drinking & cooking	2
Miscellaneous	10
Dishwashing	6
Other*	9
Total	91

11. Ground water recharge describes the process of pumping water into aquifers to maintain water volume. Purified recycled water from sewage effluent is used as recharge water. This helps maintain the ground surface level. If water is pumped out of the ground there is the possibility of settling of the surface because the space occupied by the water is now empty. There are places where the ground level has dropped by tens of feet. See photo on page 395.

13. The user was responsible for water quality prior to the 1977 Clean Water Act. The discharger is now responsible for maintaining water quality.

15. Two methods for disposal of solid wastes from industry and households are landfills and incineration. Landfills have the greater potential for water contamination because ground water can pass through them and pick up contaminants. Gas emissions from incineration can be cleaned using scrubbers to remove harmful materials, keeping them out of the environment.

17. A landfill can be made more secure with respect to water quality protection by controlling the materials placed in the landfill. The materials can be immobilized by being incorporated into concrete pellets. The storage space needs to be lined with a water proof barrier to control leaching and percolation into the ground water. Bioremediation may be used to detoxify or degrade wastes so they are no longer hazardous; this also may make landfills more secure.

19.

Household waste	Potential harmful chemicals
batteries	heavy metals
furniture polish	organic solvents
oven cleaners	caustics
paint	organic polymers
pesticides	toxic organics
weed killers	toxic organics
fertilizers	phosphates
bathroom cleaners	acids or caustics

21. Lead can be picked up from some of the solder used in joints in older copper pipe plumbing. The older solder may contain lead. This lead will dissolve in acidic drinking water. Poorly glazed ceramic mugs and pitchers may have lead pigments in the decorations. The lead in these pigments can then dissolve in acidic beverages such as tea, coffee, wine, citrus juices, carbonated beverages, etc. Ornate lead crystal contains enough lead so that similar acidic beverages can dissolve lead out from the crystal.

23. Water is vaporized from one container and condensed in another while dissolved, nonvolatile substances remain behind.

25. Settling separates high-density suspended solids such as sand, while filtration removes suspended low-density matter such as algae.

27. When chlorinated hydrocarbons and branched chain hydrocarbons are released into the environment their environmental concentrations rise because they are persistent. That means they do not decompose. They accumulate in the environment. For example, DDT, is a persistent polychlorinated hydrocarbon. It is a nonbiodegradable pesticide that is fat soluble and stored in fatty tissue of animals. To complicate matters further the DDT can also be concentrated by animal feeding habits. This can lead to increasing concentrations in animals that are higher in the food chain. Predators will concentrate these compounds in their fat tissue. DDT is implicated in interfering with bird reproduction by causing fragile eggs.

29. Ammonia, NH3(aq), and ammonium ion, NH4+(aq), can be removed from wastewater by using denitrifying bacteria. These bacteria convert the ammonia and ammonium ion to nitrogen, N2(g). The unbalanced reaction is: NH3(aq) or NH4+(aq) $\xrightarrow{\text{denitrifying bacteria}}$ N2(g)

31. Both kill bacteria by oxidizing organic compounds. Aeration depends on O2 as the oxidizing agent while chlorination uses Cl2. Chlorine gas, Cl2(g), is a good oxidizing agent. It kills pathogens by oxidizing structures in the bacterial cell. Chlorination creates the possibility of producing chlorinated organics that can be carcinogenic.

33. Mechanical pressure is used to force water molecules through a semi-permeable membrane from the salty aqueous side to the pure water side of the membrane.

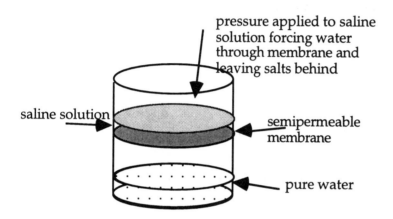

Speaking of Chemistry

Name _____

Water--Plenty of It, but of What Quality?

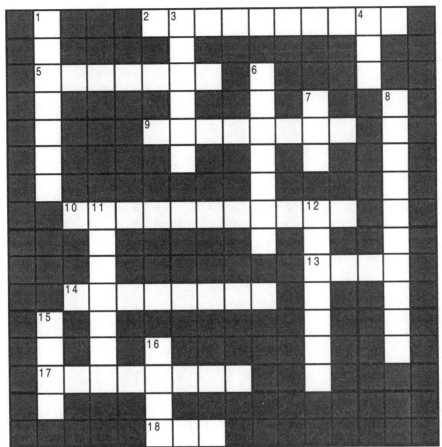

Across
2. Causes cancer
5. The flow of water from a more dilute solution through a semipermeable membrane into a more concentrated solution
9. Natural process in purification that spreads out the pollutant
10. Filtration through _____ is used to remove soluble organic compounds. Used in tertiary treatment
13. Toxic element in some paints
14. Primary water treatment includes _____ and filtration
17. Returning water to aquifers
18. A hard pesticide first used widely in 1939.

Down
1. Carcinogenic oxidation product of bromine
3. Soft water is usually _____
4. Environmental Protection Agency
6. Water suitable for drinking.
7. Biological oxygen demand, a measure of the amount of dissolved organic matter.
8. Toxic metals react with ____ groups in enzymes
11. Toxic element found naturally in soil and shrimp
12. One of the two Group IIA metal ions in hard water.
15. Resource Conservation and Recovery Act
16. Water that contains high levels of Ca^{2+}, Mg^{2+}, Fe^{3+}, Mn^{2+}

© 1999 Harcourt Brace & Company. All rights reserved.

Bridging the Gap   I   Name _____
Internet access to the Environmental Protection Agency, EPA
Non-point Pollution; Dos and Don'ts Around the Home

The role of education in bringing nonpoint-source pollution under control is extremely important. The reason for this is pragmatic: What you don't know can hurt the environment. When rain falls or snow melts, the seemingly negligible amounts of chemicals and other pollutants around your home and premises get picked up and carried via storm drains to surface waters. The ramifications include polluted drinking water, beach closings, and endangered wildlife. So what can you do to help protect surface and ground waters from so-called nonpoint-source pollution? You can start at home. Begin by taking a close look at practices around your house that might be contributing to polluted runoff; you may need to make some changes. You can become part of the solution rather than part of the problem of nonpoint-source pollution.

The EPA home page
The EPA home page has the following uniform resource locator (URL) address,

http://www.epa.gov/

The content of any displayed web page can be down loaded to disk as text or it can be printed. The underlined subject headings on the EPA home page are interactive and lead to additional pages. The Citizen Information page can be reached using the following URL.

http://www.epa.gov/epahome/Citizen.html

Remember these pages are constantly updated. The exact appearance and content of a site will change when updated.

Your responsibility is to open the following URL and use it to answer the questions on the report sheet. This site contains suggested ways to minimize water consumption and nonpoint-source pollution. Non-point source pollution is the kind of small scale pollution each isolated person creates.

http://www.epa.gov/OWOW/NPS/dosdont.html

You should also record your opinion of the EPA web page. Please explore the EPA pages and record the name and URL of a page that was interesting to you. Tell what you liked about it.

© 1999 Harcourt Brace & Company. All rights reserved.

Bridging the Gap I Name _____
Internet access to the Environmental Protection Agency, EPA
Dos and Don'ts Around the Home

1. Use this URL access the EPA page called " •Dos and Don'ts Around the Home" to do the following task.

    http://www.epa.gov/OWOW/NPS/dosdont.html

    Name two suggestions the EPA makes that you could adopt to reduce non-point source pollution.

2. The EPA gives an estimate of the volume of drinking water that one quart of motor can pollute. What is this EPA estimate?

3. Give the URL for an EPA site you found interesting.

    http://_____

    What makes it interesting to you? Would you recommend it to someone else? You can send your opinion to the EPA. They want comments.

Bridging the Gap **II** Name _____
Too Much Water
A Case Study "Waterworld"

"Waterworld" was released in 1995. It stars Kevin Costner, David Finnegan, Jeanne Tripplehorn, and Tina Majorino. The story is set in the future after the earth suffered great climate changes. The polar icecaps have melted and world is essentially covered with water. Civilization as we know it has disappeared. Surviving humans are roaming the seas struggling to find fresh water and food. Conflicting social groups fight over resources and fresh water.

Your assignment has two parts. One part is to view some of or all of "Waterworld" and develop arguments related to the limits of government action dealing with "global warming" and "water" issues. One argument should support the present policy. The other should be an argument in favor of a more intrusive policy. Write your arguments on the Concept Report Sheet.

The second part is to write your opinion about the conditions depicted in the film. Do you think it could ever happen? Why? Why not?

It is probably a good idea to view the movie with a study group. Viewing the movie with a others is probably more fun and it spreads the cost of the rental.

Specifics about the movie.
Produced by: Silver Pictures / Warner Brothers
Released in USA in 1995
Color by: Technicolor
Rating "Waterworld" has a USA "**R**" rating

Bridging the Gap **II** Name _____
Too Much Water
A Case Study "Waterworld"

1. Argument in favor of the present policy.

Argument in favor of a more active water conservation policy.

2. What is your opinion about the conditions depicted in the film? Do you think it could ever happen? What needs to be done to prevent it?

16 Air--The Precious Canopy

16.1 Air: A Source of Pure Gases

Fractionation of air for separating air into its component gases is described in detail. The boiling points for atmospheric gases are given. Physical properties of oxygen, nitrogen, and noble gases are summarized. Cryogens (ultralow temperature -183°C) and cryogenic surgery such as wart removal using $N_2(\ell)$ are described. Nitrogen fixation is mentioned.

Objectives

After studying this section a student should be able to:
- describe the fractionation of air
- explain why compressed air cools when it expands
- arrange the following gases in order of decreasing boiling point: Kr, Ar, Ne, Xe, He
- give two applications for liquid oxygen, LOX
- define cryogen and cryosurgery
- describe two hazards posed by cryogens
- explain why LOX is more hazardous than most cryogens
- tell what the nitrogen fixation does and why it is important
- give definitions for "noble gases" and an inert atmosphere

Key terms

fractionation of air	liquefaction	LOX
cryogens	cryosurgery	liquid nitrogen
liquid oxygen	nitrogen fixation	noble gas
inert atmosphere		

16.2 Doing Something About Polluted Air

A brief description of air quality problems is given. The evolution of the Clean Air Act, CAA, is summarized.

Objectives

After studying this section a student should be able to:
- name three naturally occurring atmospheric pollutants
- state the purpose of the Clean Air Act of 1970
- tell what the Clean Air Act of 1977 did to clean air standards
- describe the changes in regulations created by the Clean Air Act of 1990, 1996

Key terms

carbon monoxide	hydrogen chloride	hydrogen sulfide
carbon dioxide	smoke	chlorinated organic compounds
sulfur dioxide	oxides of nitrogen	1970 Clean Air Act
particulates	ozone	1977 Clean Air Act
sulfur	hydrocarbons	1990 Clean Air Act
volatile toxic substances		

© 1999 Harcourt Brace & Company. All rights reserved.

16.3 Air Pollutants: Particle Size Makes a Difference

Aerosols and particulates are defined. The types of matter that can be found in particulates are listed. The size of aerosol particles ranges from 1 nanometer to 10,000 nanometers, (nm = 10^{-9} m). This roughly means the 0.3 millimeter period at the end of this sentence could hold as many as 300,000 to as few as 30 aerosol particles. Adsorption and absorption are explained. Reactions and processes that can occur in an aerosol particle are illustrated. Carcinogenic and mutagenic health hazards posed by aerosols are stated. The effect of particulates on atmospheric temperature is described. Methods to remove particulates from the atmosphere such as scrubbing, electrostatic precipitation, and filtration are examined.

Objectives

After studying this section a student should be able to:
- give definitions for particulates and aerosols
- state the range of diameters for aerosol particles
- describe how aerosol particles can pick up hazardous substances and concentrate them
- explain how aerosol particles can alter the temperature of the Earth
- describe the connection between particulates and volcanic eruptions
- tell how particulates and aerosols are naturally removed from the atmosphere
- name three ways to remove particulates and aerosol particles from air
- sketch diagrams showing how adsorption and absorption differ

Key terms

particulates	aerosols	nanometer
mutagenic compounds	carcinogenic compounds	adsorb
soot	suspended particulate	gravitational settling
rain and snow	filtration	physical methods
scrubbing	electrostatic precipitation	

16.4 Smog

A brief history of smog is given. Thermal inversions trap pollutants near the ground and contribute to smog production. Chemically reducing smog containing soot, fly ash, and partially oxidized organics is described. The health effects of chemically reducing smog are summarized. The origins of photochemical smog containing ozone, nitrogen oxides, organic peroxides, etc. are described. Primary pollutants, secondary pollutants, free radicals, and the sources of atmospheric hydrocarbons are discussed.

Objectives

After studying this section a student should be able to:
- give the definitions for smog, inversion layer, photochemical smog, industrial smog
- name the two general types of smog
- name the sources of reducing smog
- describe the effects of reducing smog
- state the definition of a free radical and give an example
- give definitions for primary and secondary pollutants
- write the equation for the reaction between NO_2 and UV light to form O atoms and NO

Key terms

smog	thermal inversion	chemically reducing type
industrial smog	sulfur dioxide, SO_2	sulfuric acid, H_2SO_4
emphysema	asthma	respiratory diseases
photochemical smog	nitrogen dioxide, NO_2	chemically oxidizing type
ozonated hydrocarbons	unreacted hydrocarbons	
free radical	primary pollutants	secondary pollutants

16.5 Nitrogen Oxides

The many oxides of nitrogen are described and the general formula NO_x is given. The natural sources of nitrogen oxides are named. Some reactions of NO and NO_2 are illustrated. Photodissociation is defined and illustrated with the reaction

$$NO_{2(g)} + h\nu \longrightarrow NO_{(g)} + O_{(g)}$$

The wavelengths of effective UV light are given. The reaction between water and NO_2 is given and related to aerosol droplet stability. Formation of secondary pollutants is explained.

Objectives
After studying this section a student should be able to:
- tell how NO is formed naturally and write the equation for the reaction
- identify a natural source of N_2O
- describe how NO behaves in the atmosphere
- describe the physiological effects of NO_2, nitrogen dioxide
- state the definition for photodissociation
- tell how ozone is produced from NO_2, UV light, and O_2 (include reactions)
- write the reaction for the combination of water and nitrogen dioxide and tell what the products do to the stability of aerosol drops
- tell what a secondary pollutant is

Key terms
oxides of nitrogen, NO_x	dinitrogen oxide, N_2O	nitric oxide, NO
nitrogen dioxide, NO_2	parts per billion, ppb	haze
combustion temperature	bronchioconstriction	photodissociation

16.6 Ozone and Its Role in Air Pollution

The formula for ozone, $O_{3(g)}$, is given and its sources are identified. The difference between "good" and "bad" ozone is explained. Ozone as a pollutant in the troposphere is discussed. The new 1997 EPA standard for pollutant ozone is stated as 0.08 ppm and discussed in terms of FEV_1 values. The relationship between O_3 and NO_x is explained.

Objectives
After studying this section a student should be able to:
- write the formula for ozone
- tell the difference between "good" and "bad" ozone
- tell why the EPA pollutant level for ozone may be too low
- give the definition for forced air volume, FEV_1
- describe what ozone does to the FEV_1 of children
- explain why ozone concentrations and NO_x concentrations are related

Key terms
ozone, O_3	"good" ozone	"bad" ozone
stratosphere	ultraviolet radiation	forced expiratory volume, FEV_1
NO_x emissions		

16.7 Hydrocarbons and Air Pollution

Natural sources of hydrocarbons like decaying organic matter, deciduous trees, coniferous trees, plants, etc. are identified. Human sources of hydrocarbons such as polynuclear aromatic hydrocarbons (PAH) and the carcinogenic (BAP) are discussed. Efforts to control human sources of hydrocarbon emissions are described. The effect of catalytic converters on automobile hydrocarbon emissions is discussed.

© 1999 Harcourt Brace & Company. All rights reserved.

Objectives
After studying this section a student should be able to:
- name natural sources of hydrocarbon emissions
- tell what percent of atmospheric hydrocarbons come from natural sources
- explain why hydrocarbons like PAN and BAP are of special concern
- describe how automobile hydrocarbon emission rates have changed since 1960
- explain why hydroxyl radicals are a pollution problem

Key terms

hydrocarbons	methane gas, CH_4	alkanes
alkenes	alkynes	catalytic converters
PAH	BAP	benzo(α)pyrene
carcinogen	hydroxyl radicals	photodecomposition
polynuclear aromatic hydrocarbons		

16.8 Sulfur Dioxide: A Major Primary Pollutant

The natural (volcanic eruptions) and human sources (burning of sulfur bearing matter) of sulfur dioxide, SO_2, are identified. The effect of SO_2 on the Earth's temperature is explained in terms of stabilized aerosol droplets that reflect sunlight. The connection between sulfur bearing fossil fuels and atmospheric SO_2 is discussed. Two methods for the removal of SO_2 from exhaust gases are described. The reduction of ground level SO_2 concentrations using tall stacks to dilute exhaust gases is discussed. The reactions of SO_2 to form SO_3 and sulfuric acid are described.

Objectives
After studying this section a student should be able to:
- give the formula for sulfur dioxide
- explain how SO_2 in the atmosphere can alter the Earth's temperature
- identify the human sources of SO_2
- describe the EPA requirements proposed for the year 2000 for power plant SO_2 emissions
- tell how SO_2 is removed from power plant exhaust gases
- explain why power plants that burn sulfur bearing fossil fuels have tall stacks
- write the equation for the reaction of $SO_3(g)$ and water to form $H_2SO_4(aq)$

Key terms

sulfur dioxide, SO_2	aerosol droplets	scattering of sunlight
pyrite, FeS_2	calcium sulfite	hydrogen sulfide, H_2S
high-sulfur coal	electrostatic precipitator	tall stacks, 500 to 1000 feet

16.9 Acid Rain

Normal rain fall is not neutral at pH = 7. Normal rain contains dissolved CO_2 and has a pH of 5.6. Acid rain or acid deposition is defined as rainfall with a pH lower than 5.6. The acidic gases SO_2 and NO_x are identified as contributors to acid rain. The source locations of acidic oxide gases and the sites that receive the fallout are mapped for the Northeastern United States. The pH values for regions in the Northeastern United States are illustrated. The maps show the international nature of the acid rain problem. Acid rain is described as responsible for producing "dead", fishless ponds and lakes. The pH levels that various species of fish can tolerate are given. The damage acid rain does to forests and individual trees is described. The effects of acid rain and other atmospheric pollutants on stone and metal structures are discussed. Examples of damage to historic structures caused by acid rain are described. Efforts to prevent and to correct the damage done by acid rain are summarized. The international aspect of acid rain is described as a barrier to solving the problem.

Objectives
After studying this section a student should be able to:
- tell why "pure" rain water is acidic and give the normal pH for rainwater
- give the pH for neutral water
- give the definition for acid rain
- explain how gases like CO_2, O_2, and NO_x generate acid rain
- describe the problems acid rain creates for fish in ponds and small lakes
- tell how acid rain damages trees and forests
- describe how acid rain damages stone and metal structures
- explain why acid rain is a national and an international problem
- explain what the term "fallout" refers to when discussing acid rain

Key terms
acid rain	pH below 5.6	nitric acid, HNO_3
nitrous acid, HNO_2	sulfuric acid, H_2SO_4	sulfurous acid, H_2SO_3
"dead" ponds and lakes	dying or dead forests	acid rain tree damage

16.10 Carbon Monoxide
The natural and human sources of carbon monoxide, CO, are described. Natural sources of CO are described as ten times greater than human sources. The major source of human generated CO is automobile exhaust. The constant level of atmospheric CO is currently not explainable. Carbon monoxide toxicity and carbon monoxide poisoning are explained. The reaction of CO and hemoglobin, and the treatment for carbon monoxide poisoning are is discussed.

Objectives
After studying this section a student should be able to:
- name a natural source of CO
- name the biggest artificial source of CO
- tell how the amounts of natural and human generated CO compare
- write the equation for the reaction between insufficient oxygen and coal to yield CO
- explain why automobile catalytic converters play a role in limiting CO emissions
- explain why carbon monoxide is poisonous

Key terms
carbon monoxide	natural sources	oxidation of decaying organic matter
industrial sources	automotive sources	gasoline engines
catalytic converters	hemoglobin	carbon monoxide poisoning

16.11 Chlorofluorocarbons and the Ozone Layer
Chlorofluorocarbons (CFCs) are defined and examples are illustrated. Uses for CFCs are described along with a brief history of their development. The photochemical reaction of CFC-11 is used as an example of how CFCs can be a source of Cl• and ClO• radicals. The role of ozone as a UV absorber is described. Ozone depletors are identified and the steps in the chain reaction for ozone depletion are discussed. Natural ozone depletors are discussed. The Montreal Protocol and efforts to control CFC emissions are summarized. Alternatives to CFCs are discussed and the economic costs of banning CFCs are reviewed. The weakness of the carbon-chlorine bond is also explained. (Dots indicate unpaired electrons.)

CFC-11   CFC-12   CFC-11 + UV light →   + Cl•

Objectives
After studying this section a student should be able to:
- write the formula for ozone
- give the definition for CFCs and an example
- explain why CFCs were used as refrigerants, degreasers, etc.
- describe how photodissociation makes CFCs a threat to stratospheric ozone
- write the reaction for the photodissociation of CFC-11
- tell why stratospheric ozone is "good"
- explain what an "ozone hole" means
- describe the role of ozone in absorbing ultraviolet light
- describe the health problems that will result from ozone depletion
- summarize the chronology of the political actions to restrict CFCs
- describe compounds such as perfluorobutane that are being used in place of CFCs
- tell why naturally occurring CH_3I is able to be an ozone depletor
- describe the likely future for CFC usage

Key terms

CFCs	chlorofluorocarbons	halogenated hydrocarbons
refrigerant gas	degreasers	propellant
CFC-11	CFC-12	ozone-depleting substances
stratospheric ozone	ultraviolet light	black-market
200-300 nanometer range	photodissociation	chlorine oxide, ClO•
Montreal Protocol	ozone hole	hydrochlorofluorocarbons
retrofitted	HCFC-134a	HCFC-141b
halons	bromodifluoromethane	bromine oxide radicals, BrO•
perfluorobutane	"per-"	chlorotetrafluoroethane

16.12 Carbon Dioxide and the Greenhouse Effect

The trends in atmospheric CO_2 concentrations are graphed for the period 1958 to 1990 and discussed. The link between increased fossil fuel use and CO_2 concentrations is described. The energy balance between solar energy received by the earth and energy radiated into space is described. The changes in CO_2 concentrations are related to the greenhouse effect. The greenhouse effect is illustrated and the process is explained. Greenhouse gases like O_3, CH_4, H_2O, and CO_2 are identified. Projected increases in atmospheric temperature are linked to CO_2 concentrations above 360 ppm. Political difficulties in solving the global warming problem are discussed. International efforts to control CO_2 emissions to minimize global warming are described.

Objectives
After studying this section a student should be able to:
- describe the trend in atmospheric CO_2 levels since 1958
- explain why CO_2 levels have increased
- tell how CO_2 concentrations influence atmospheric temperature
- give the definition for a greenhouse gas
- explain what the greenhouse effect is
- name four greenhouse gases
- give the definition for global warming
- tell how greenhouse gases alter the atmospheric temperature
- explain why controlling the greenhouse effect is an international problem
- describe the political problems associated with controlling global warming

summarize the United Nations projections for greenhouse gas concentrations

Key terms

fossil-fuel	carbon dioxide, CO_2	global concentration of CO_2
cut-and-burn	OPEC	solar radiation
visible light	reradiate	infrared
water vapor	ozone	methane, CH_4
greenhouse effect	greenhouse gases	atmospheric temperature
absorbing blanket	global warming	global industrial emissions

16.13 Industrial Air Pollution

Industrial chemical releases are described and compared to natural chemical releases. The need for industrial disclosure of chemical releases is discussed. The five states with the greatest amount of chemicals released are identified. The United States EPA requirement for industrial release reports and Community-Right-to-Know regulations are discussed. The EPA hotline for information about reported chemical releases is given. 1-800-535-0202.

Objectives

After studying this section a student should be able to:
- describe the effects of the chemical release that occurred in Bhopal, India
- tell how the Bhopal incident influenced the EPA release reports
- give the approximate number of compounds that industry must file release reports for
- explain the reasoning behind the Community-Right-to-Know regulations
- give the EPA toll free number that gives information about reported chemical releases

Key terms

industrial releases	summary of annual releases	Superfund
release report	Community Right To Know	solvents
NO_x	SO_2	CO_2
CFC's	metal particulates	acid vapors
unreacted monomers	EPA	1-800-535-0202

16.14 Indoor Air Pollution

The EPA studies of indoor air are described. The sources of indoor pollutants are identified. The move to make homes energy efficient and air tight is identified as one reason for indoor pollution. A schematic of a typical home is used to show the types and sources of indoor pollutants. Indoor pollution in nonsmoker homes is compared with pollution in the homes of smokers. The "sick building syndrome" is described.

Objectives

After studying this section a student should be able to:
- give reasons why indoor pollution is more of a problem than in previous years
- name typical indoor pollutants and identify their sources
- tell how indoor pollutants differ in a nonsmoker's home from a smoker's home
- explain what is meant by the term "sick building syndrome"

Key terms

indoor air pollution	energy-efficient homes	carcinogen
benzene	tobacco smoke	radon-222
perchloroethylene	formaldehyde	para-dichlorobenzene
fungi	carbon monoxide	methylene chloride
nitrogen oxides	chloroform	sick building syndrome

© 1999 Harcourt Brace & Company. All rights reserved.

Additional readings

Beardsley, Tim. "Add Ozone to the Global Warming Equation." Scientific American Mar. 1992: 29.

Calvert, Jack G., et al. "Achieving Acceptable Air Quality: Some Reflections on Controlling Vehicle Emissions" Science July 2, 1993: 45.

Greene, David L. and K. G. Duleep. "Costs and Benefits of Automotive Fuel Economy Improvement: A Partial Analysis" Transportation Research May 1993: 217.

Matthews, Samuel W., and James A. Sugar. "Under the Sun: Is Our World Warming?" National Geographic Oct. 1990: 66.

Mohnen, Volker A. "The Challenge of Acid Rain" Scientific American Aug. 1988: 30.

Viggiano, A.A., et al. "Ozone Destruction by Chlorine; the Impracticality of Mitigation Through Ion Chemistry" Science Jan. 6, 1995: 82.

Washington, Warren M. "Where's the Heat?" Natural History Mar. 1990: 67.

Zimmer, Carl. "Unintended Consequences" Discover Mar. 1995: 32.

Answers to Odd Numbered Questions for Review and Thought

1.
   a. Polluted air is air that contains unwanted and harmful substances.
   b. Aerosols are mixtures of water droplets and particulates with diameters in the range of 1 nm - 10,000 nm.
   c. Photochemical smog is chemically oxidizing and formed by the action of sunlight on combustion engine exhaust.
   d. Secondary pollutants are not directly emitted but are formed by reactions between pollutants and other compounds in the air.

3.
   a. Acid rain is rainwater with a pH lower than 5.6.
   b. CFC is the abbreviation for chlorofluorocarbons such as CFC-11, CCl_3F, and CFC-12, CCl_2F_2.
   c. The ozone hole is a region the stratosphere that has lower than normal ozone concentration. This lowered concentration level results in a more UV light reaching the earth's surface.
   d. A greenhouse gas is a molecule that absorbs infrared light and radiates the energy back to the atmosphere.
   e. Global warming refers to a worldwide increase in atmospheric temperature.
   f. A cryogen is a liquified gas that exists at extremely low temperatures like liquid nitrogen at -196°C (77 kelvins).

5. Clean Air Act has the abbreviation CAA. The first act was passed in 1970 and originally controlled air pollution from cars and industry. Amendments in 1977 imposed stricter auto emission standards. The 1990 CAA extends to manufacturing and commercial activity. This most recent version of the CAA regulates particulates, ozone, carbon monoxide, oxides of nitrogen and sulfur, carbon dioxide, and substances that would deplete stratospheric ozone. A newer version of the CAA was passed in 1996.

7. Industrial smog is different from photochemical smog because it is chemically reducing while photochemical smog is chemically oxidizing. Industrial smog contains sulfur dioxide mixed with soot, fly ash, smoke, and partially oxidized organic compounds. Photochemical smog is essentially free of SO_2, but contains ozone, ozonated hydrocarbons, organic peroxide compounds, nitrogen oxides, and unreacted hydrocarbons.

9. Ozone is formed in the stratosphere when UV light breaks up O_2 to produce oxygen atoms; these react with additional O_2 to form O_3, ozone.

11. Of the nitrogen oxides "NO_x" in the atmosphere, 97% come from natural sources. Lightning strikes during electrical storms produce NO. Some bacteria produce N_2O. Nitric oxide, NO, is so reactive that it combines with O_2 in the atmosphere to produce NO_2. About 3% of the atmospheric nitrogen oxides come from human activity such as combustion in automobile engines.

13. Volcanic eruptions can contribute to global cooling because the eruption will throw dust particles into the air. These dust particles scatter and reflect sunlight into space, so the solar energy never reaches the earth's surface. This will decrease the amount of energy striking the earth and decrease the atmospheric temperature.

© 1999 Harcourt Brace & Company. All rights reserved.

15. Hydrocarbons are released into the atmosphere by natural sources and human sources. Hydrocarbons are put into the atmosphere by living plants such as deciduous trees, by the decay of dead plants and animals, and by excrement from insects and animals. Human activity introduces hydrocarbons when organic solvents are used, for example in the handling of petroleum products, etc. Generally only human activities are within our control.

17. Nitrogen dioxide plays a role in the formation of ozone in the troposphere.

19. Nitrogen dioxide (NO_2) will dissociate to form an oxygen atom and nitric oxide, if it is excited by a sufficiently energetic photon of UV light. The O atom forms O_3 by reaction with O_2.
    $$NO_2(g) + h\nu \longrightarrow NO(g) + O(g)$$
    $$O + O_2 \longrightarrow O_3$$

21. All three are oxidizing agents. Ozone, O_3; sulfur dioxide, SO_3; and nitrogen dioxide, NO_2; cause lung damage.

23. Pure oxygen is removed from the air by the process of liquefaction or fractionation of air. The process depends on the fact that each gas in has a definite and distinct boiling temperature. The other principle is that the expansion of a gas draws energy from the gas and cools it. This means that when a gas is compressed and then allowed to expand its temperature will fall.

    The air is precooled below 0°C = 273 kelvins to remove water vapor as ice. The temperature is decreased to less than -78°C or 194 K to remove CO_2. The dry air, free of CO_2, is compressed to more than 100 atmospheres. This compression heats the air because energy is added to compress it. The air is cooled to room temperature. The room temperature air is allowed to expand and cool. This compression expansion cycle is repeated until all of the air is liquefied. The liquid air is allowed to warm and each gas in the mixture will "boil" off at its own boiling point.

25. Particulate air pollutants can cause lung diseases, cancer and mutagenic effects.

27. Automobile emissions in grams/ mile (g/mile)

1960 (no catalytic converters)	g/mile	1993 (catalytic converters)	g/mile	1995 (estimated)	g/mile
HC	10.6	HC	0.41	HC	0.41
CO	84.0	CO	3.4	CO	3.4
NO_x	4.1	NO_x	1.0	NO_x	1.0

    All three categories of emissions decreased on a per mile basis because of catalytic converters. Lead emissions decreased because unleaded gasolines are now used. These data show what happens per mile of automobile use. What these do not show is how the totals compare because of changes in the number of miles driven. The cars are cleaner, but increases in total miles driven could increase actual pollution.

29. Oil burning electric fuel plants generate SO_2. The amount of SO_2 emitted can be reduced by either using low sulfur fuel oil or by passing plant exhaust gas through molten sodium carbonate to form sodium sulfite.

31. Rain tends to be acidic because it dissolves CO_2 to make H_2CO_3. This normally creates a solution with a pH of 5.6. Human sources put SO_2 and NO_x into the atmosphere. They react with rainfall to yield nitric acid, sulfuric and sulfurous acid. The pH of these mixtures will be lower than the natural 5.6. This precipitation is called acid rain. Values have been observed in the pH range of 4.0 to 4.5. Isolated storms have yielded rain with a pH of 1.5.

33. The reaction between carbon and oxygen when there is insufficient oxygen is
    2 C + O_2 ----> 2 CO   The product is carbon monoxide.

35. CFCs are linked to depletion of ozone concentrations in the stratosphere. These regions of lowered ozone concentration are relative ozone "holes". These ozone holes lead to increased ultraviolet light levels at sea level resulting in greater rates of skin cancer. Some unforeseen effects on algae, plants and animals that could upset the ecological balance are also causes for worry.

37. Ultraviolet light can cause the break up of oxygen molecules. O_2 + hv ------> 2 O
    The oxygen atoms formed by the dissociation of O_2 can form ozone in the following reaction. O_2 + O $\longrightarrow$ O_3

39. Automobile air conditioners primarily used CFC-11, CCl_3F. CFCs can circulate in the atmosphere and reach the stratosphere. There ultraviolet light can break the carbon chlorine bond to produce Cl atoms. These Cl atoms can react with O_3 to deplete ozone concentrations in the stratosphere. The free radicals each have an unpaired electron and are extremely reactive. The ultraviolet light is indicated by the hv symbol. CFCs are no longer used in the air conditioners of new automobiles in order to avoid contributing to these reactions and to the ozone depletion problem.

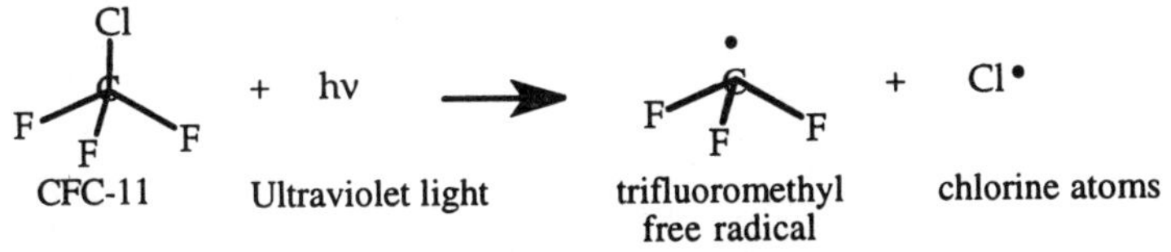

    CFC-11 + Ultraviolet light $\longrightarrow$ trifluoromethyl free radical + Cl•

Speaking of Chemistry

Air: The Precious Canopy

Name _____

Across
2. Ozone in stratosphere
4. Replacements for CFCs
7. Organization of Petroleum Exporting Countries
9. Process of attaching to a surface
11. Reactive species with unpaired electrons, Cl
12. Benzo(a)pyrene
15. Forced expiratory volume
16. Compounds like NO⊕—⊗, NO₂⊕⊗⊗
19. Compounds that may cause mutations
21. Fog and ____ are examples of aerosols
22. Smoke, dust and ____ are suspended particulates
23. Reacts with water to form sulfuric acid, H_2SO_4, and sulfurous acid, H_2SO_3.

Down
1. Polynuclear aromatic hydrocarbons
2. Increased atmospheric temperature
3. Most abundant gas in the atmosphere
5. Compounds linked to ozone depletion, chlorofluorocarbons
6. Clean Air Act
8. Ozone at ground level
10. Compounds used in fire extinguishers like CF_2BrCl, 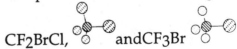 and CF_3Br
13. UV absorbing gas
14. Effect caused by gases like CO_2, CH_4
17. Environmental Protection Agency
18. Regions in stratosphere with thinner than normal ozone levels
20. Light absorbed by ozone layer

© 1999 Harcourt Brace & Company. All rights reserved.

Bridging the Gap   I   Name_____
Internet link to the Center for Atmospheric Science

The Earth's Atmosphere and the ozone Layer
Cambridge University in Cambridge England is home for the Centre for Atmospheric Science. The Centre maintains and excellent website on the properties of the atmosphere. The site includes a "tour" of atmosphere related topics. The page for the various activities is readily accessible using the following uniform resource locator (URL) web site address.

http://www.atm.ch.cam.ac.uk/tour/atmosphere.html

The Earth's Atmosphere: How high does it go?
You can scroll down the table of contents page and click on the active line named

1. What are the five "layers" that make up the earth's atmosphere?

2. How high, in kilometers, does the atmosphere reach?

3. How high up, in kilometers, is the stratosphere?

The History behind the Ozone Hole and the Threat it Poses
The first observations of ozone depletion in the lower stratosphere were made in the 1970s by a research group from the British Antarctic Survey. Use this URL to answer the questions below.
    http://www.atm.ch.cam.ac.uk/tour/part1.html

4. What is ozone? Write its formula.

5. How is ozone formed in the stratosphere? Write the two reactions in the process.

6. What good does ozone in the stratosphere do for humans on the ground?

7. What is the ozone hole? Why is the depletion of the ozone layer a concern for humans?
    http://www.atm.ch.cam.ac.uk/tour/part3.html

© 1999 Harcourt Brace & Company. All rights reserved.

Bridging the Gap II Name _____
Global Warming and Individual Responsibility

The global warming problem is tied to greenhouse gases. Natural greenhouse gases are beyond human control. The artificial sources of greenhouse gases are controllable, at least in theory.

The social and political problems of reducing greenhouse gas emissions are tremendous. You will probably read in Chapter 17 that the global population is projected to be about 11.5 billion people by the year 2050. This projection means that the global population will essentially double in about 50 years.

There are studies that say the temperature of the atmosphere might remain constant if CO_2 concentrations in the atmosphere can be kept at 360 ppm. Do you think a doubling of the Earth's population can happen without exceeding this CO_2 concentration? Your assignment is to give an answer to this question and justify your answer. Your second assignment is to meet with other members of your class and discuss what will happen to CO_2 concentrations if people continue their present patterns of fossil fuel use. Your discussion should be aimed at developing two suggestions for changes in individual behavior that will help keep CO_2 concentrations from rising.

Use this page to record your suggestions and summarize your group's discussion.

17 Feeding the World

17.1 World Population Growth
International conferences on global population are listed. The current global population is given as approximately 5.9 billion. The rate of global population growth is described. The year 2050 is given as the time when the world population will be double what it is now. Estimates are given for the number of people currently living in poverty.

Objectives
After studying this section a student should be able to:
- state the approximate global population in 1995
- give the projected global population in 2050
- give an estimate of the percentage of the global population that lives in poverty
- explain why there is concern about global population growth and agricultural productivity

Key terms
- Worldwatch Institute
- Earth Summit
- agricultural productivity
- agrichemicals
- world population
- rate of growth of world's population
- United Nations Conference on Environment and Development
- United Nations International Conference on Population and Development

17.2 What is Soil?
The structure of soil is described, and the four components of soil are listed. The function of humus is described. The structure of soil is illustrated and labeled. Definitions are given for horizons, humus, friable soil, topsoil, subsoil, separates, and loam. The differences between normal air and air mixed in soil are explained. The effect of CO_2 levels in soil on its pH is described. Sweet and sour soils are defined. The percolation process for water movement in soil is described. Leaching is defined; selective leaching of soil and its effect on metal ion and salt concentrations are explained.

Objectives
After studying this section a student should be able to:
- name the four components of soil
- give definitions for humus, horizons, topsoil, subsoil, loam
- describe how oxidation of organic matter in soil alters it's pH
- describe the difference between sweet soil and sour soil in terms of pH
- describe the three ways water is stored in soil
- describe the percolation process for water movement through soil
- explain how leaching of soil affects concentrations of salts of Group IA and IIA metals
- describe how leaching of soil affects the concentrations of salts of Group IIIA
- explain what is meant by selective leaching, tell how it affects soil pH
- name two processes that influence soil pH
- sketch and label a diagram of soil showing horizons

Key terms

soil	mineral particles	organic matter
soil structure	friable	humus
topsoil	subsoil	horizons
clay	silt	sand
gravel	loam	groundwater
air in soil	normal dry air	sweet soil
acidic soil	sour soil	pores
absorbed	adsorbed	salts of Group IA & IIA metals
transpire	waterlogged soil	salts of Group IIIA metals
percolation	selective leaching	salts of transition metals
leaching effects		alkaline

17.3 Nutrients

Elemental nutrients are classified as nonmineral and mineral. The nonmineral nutrients are identified as carbon, hydrogen and oxygen and can be obtained from the atmosphere. Mineral nutrients are subdivided into primary, secondary and micronutrients. They are available to plants as solutes through the plant roots. Primary nutrients are identified as nitrogen, potassium, and phosphorus. The water soluble forms of primary nutrients are described. The effect of pH on the solubility of phosphate is described. The solubilities of secondary nutrients, calcium and magnesium, are discussed. Iron is defined as a micronutrient and the effect of pH on its solubility is explained.

Objectives

After studying this section a student should be able to:
  explain how nonmineral and mineral nutrients differ
  name the three nonmineral nutrients
  explain why mineral nutrients must be water soluble to be used by a plant
  give the definition for nitrogen fixation
  describe the role of nitrogenase in nitrogen fixation by legumes
  describe how pH influences the solubility of phosphate
  describe the symptoms of chlorosis and tell what causes it

Key terms

elemental nutrients	nonmineral nutrients	mineral nutrients
primary nutrients	secondary nutrients	micronutrients
nitrogen fixation	nitrogenase	soluble mineral
nitric acid, HNO_3	$H_2PO_4^-$	dihydrogen phosphate ion
nitrate salts	HPO_4^{2-}	monohydrogen phosphate ion
K^+	Ca^{2+} and Mg^{2+}	trivalent phosphate ion, PO_4^{3-}
chlorophyll	chlorosis	complex ions

17.4 Fertilizers Supplement Natural Soils

A brief history of fertilizer development is given. The usable life of land under a "slash-burn-cultivate" cycle is estimated to be only one to five years. The costs of applying chemical fertilizers worldwide are discussed. Productivity of chemically fertilized land is compared with unfertilized land. The labeling system for chemical fertilizers is explained. Examples are given for calculations of weights of N, P and K for labeled fertilizers. Quick release and slow release fertilizers are defined. Nitrogen fertilizers like urea, ammonia, and ammonium nitrate are identified. A possible solution to the misuse of ammonium nitrate by terrorists is described. Production and use of superphosphate fertilizer is discussed.

Objectives
After studying this section a student should be able to:
- tell how long farmland used in the slash-burn-cultivate cycle can produce crops
- state definitions for straight, complete, and mixed fertilizers
- tell why ammonia, urea, and ammonium nitrate are applied to soil
- tell what a fertilizer grade label indicates about nutrient concentrations
- name the compound made in the Haber process
- use a fertilizer label to calculate
    - the weight of nitrogen present as N,
    - the weight of phosphorus present as P_2O_5 and
    - the weight of potassium as K_2O
- explain why "superphosphate" is used as a fertilizer instead of $Ca_3(PO_4)_2$

Key terms
slash-burn-cultivate cycle	legumes in crop rotation	mixed fertilizer
straight fertilizer	complete fertilizer	grade
phosphorus as P_2O_5	potassium as K_2O	potash
quick-release fertilizer	slow-release fertilizer	"phosphate"
anhydrous ammonia	Haber process	urea
explosives	"liquid nitrogen"	ammonium nitrate, NH_4NO_3
$Ca_3(PO_4)_2$	phosphate rock	superphosphate

17.5 Protecting Food Crops

Pesticides, insecticides, herbicides, and fungicides are discussed. The economic impact of these compounds is explained. A summary of the historical development of insecticides is given. The development of DDT is discussed. Benefits and problems with persistent chlorinated hydrocarbon pesticides are explained. The LD_{50} values for various pesticides are compared. Environmentally friendly pesticides are described. Tillage is defined. Selective and nonselective herbicides, and the enzyme inhibiting herbicides like glyphosate are discussed. The widely used herbicide 2,4-D is described. The reasons for banning 2,4,5-T are given. Pre-emergent herbicides like paraquat are described. Traditional fungicides like copper and sulfur compounds are identified. New fungicides such as Thiram are described.

2, 4, 5-trichlorophenoxyacetic acid
2, 4, 5-T

2, 4-dichlorophenoxyacetic acid
2, 4-D

DDT

Thiram

© 1999 Harcourt Brace & Company. All rights reserved.

Objectives

After studying this section a student should be able to:
- give the definitions for herbicides, fungicides, insecticides, and pesticides
- explain what is meant by "persistent" and "nonpersistent" pesticides
- give three reasons for a decline in the use of pesticides in the US
- describe the history of DDT and tell why it is banned in the United States
- explain what makes carbamate insecticides attractive
- describe the reasons why insecticides are used even though they may pose other problems
- explain what it means when a substance is said to have a low LD_{50}
- tell why 2, 4, 5-T is banned by the EPA
- give the definition for tillage and explain why herbicides make it less necessary
- give an example of a common herbicide
- describe how glyphosate functions as a herbicide
- name a pre-emergent herbicide
- explain why fungicides are necessary
- name an "old" fungicide and a "new" fungicide

Key terms

herbicides	biological half-life	World Health Organization
fungicides	chlorinated hydrocarbons	organophosphorus compounds
insecticides	carbamates	toxicity
persistent pesticides	DDT	fat soluble
biodegradable	LD_{50}	Malathion
selective herbicide	selective insecticide	minimum tillage
growth hormones	tillage	Dioxin
triazines	Agent Orange	dithiocarbamates
glyphosate	Round-up	inhibiting plant enzymes
sulfonylureas	Oust	essential amino acids
Paraquat	pre-emergent herbicide	mildew
copper compounds	mercury compounds	sulfur
2,4-dichlorophenoxyacetic acid		2,4-D
2,4,5-trichlorophenoxyacetic acid		2,4,5-T

17.6 Sustainable Agriculture

The magnitude of the topsoil erosion problem is described. Deficiencies in farming practices are identified. Pest resistance to pesticides and soil compacting are two problems resulting from poor practices. Sustainable and organic farming are discussed as alternatives to poor farm practices. National Research Council recommendations for new farming practices are given. Integrated pest management, IPM, is described as a way to reduce pesticide usage. Natural insecticides, insect pheromones, disease-resistant crops, and insect pest predators are discussed as parts of the IPM program. Natural insecticides like pyrethrum and oil extracted from seeds of the neem tree are described. Natural selective insecticides based on bacterial toxins are discussed.

Objectives

After studying this section a student should be able to:
- give approximate values for regional soil erosion in tons per acre
- explain what soil compacting does to soil fertility
- describe the main aspects of organic farming
- state the six proposals in the 1989 National Research Council report
- describe features of the integrated pest management system
- explain why natural insecticides are desirable

describe the benefits of pyrethrum and pyrethrin insecticides
explain why pyrethrum must be mixed with other insecticides to enhance its effectiveness
describe the advantages offered by insecticides based on *Bacillus thuringiensis* (*Bt*)
describe the advantages of insecticides based on neem tree oil
explain why rotating crops helps fight insect infestations

Key terms

organic farming	alternative agriculture	sustainable agriculture
natural fertilizers	green manure	biological pest controls
insect pheromones	natural insecticides	integrated pest management
pyrethrum	bacterial toxins	*Bt* strains
Bt toxins	biopesticides	azadirachtin
neem-tree seed oil		

17.7 Agricultural Genetic Engineering

Transgenic crops are defined. The development time for natural breeding methods is compared with the development time for genetically engineered plants. The ability to engineer plants with specific traits is described. Tomatoes resistant to tobacco mosaic virus and plants that produce their own insecticides are examples of successful genetically engineered crops.

Objectives

After studying this section a student should be able to:

give the definition for transgenic plants or crops
tell why developing genetically engineered crops is quicker than traditional breeding
tell why people oppose genetic engineering experiments with plants
give an example of a successful genetically engineered crop

Key terms

transgenic crops	genetically engineered	natural breeding
FlavrSavr tomato		

Additional Readings

Christensen, Damaris. "Microorganisms Used to Fight Fruit Rot" Science News June 4, 1994: 359.

Cooper, Mary H. " Regulating Pesticides: Do Americans Need More Protection From Toxic Chemicals?" CG Researcher Jan. 28, 1994: 75.

Hanson, David. "Pesticides and Food Safety: Administration Proposes Broad Reforms" Chemical and Engineering News Jan. 28, 1994: 6.

Newman, Alan. "Would Less Frequent Pesticide Monitoring be Better?" Environmental Science & Technology Feb. 1995: 70.

"Pest Wars" National Geographic April 1992: 13.

Raloff, Janet. "Berry Scent Defends Fruit From Fungus" Science Aug. 7, 1993: 84.

Safrany, David R. "Nitrogen Fixation" Scientific American Oct. 1974: 64.

Schneider, Dietrich. "The Sex-attractant Receptor Of Moths" Scientific American July 1974: 28.

Sheldon, Richard P. "Phosphate Rock" Scientific American June 1982: 5.

© 1999 Harcourt Brace & Company. All rights reserved.

Answers for Odd Numbered Questions for Review and Thought

1. The earth's carrying capacity depends on the amount of productive land, agricultural practices, biotechnology advances, the rate of degradation of farmland global pollution, the amount of water for irrigation, and the level of an acceptable quality of life.

3. Soil is a mixture of mineral particles, organic matter, water, and air.

5. Soil has a structure made up of a series of layers called horizons which are loosely packed and permeable near the surface and gradually change to impermeable solid rock. The upper most horizon or topsoil is a permeable mixture of living organisms and humus. The next horizon down is the subsoil which is permeable or friable clay in the upper layer and stiff clay below. The deepest horizon is the substratum which has an upper level of soft rock and a lower layer of solid rock.

7. Limestone is $CaCO_3$. It is a basic compound and reacts with acids, H^+, to produce bicarbonate ion and Ca^{2+} ion. Limestone will neutralize acids in the soil. Adding a base such as limestone to soil causes the pH to go up. This corresponds to a sweet soil. Low pH values below 7 indicate acidic conditions, 7 indicates a neutral condition and values above 7 indicate basic.

9. The three nonmineral nutrients that plants need for healthy growth are carbon, hydrogen and oxygen. Plants get carbon from carbon dioxide in the atmosphere. They get hydrogen and oxygen from water.

11. a. Metal ions from Group IA and Group IIA are more easily leached. These ions are generally more water soluble than the Group IIIA metal ions. Leaching occurs when water removes the metal ions from the soil.
    b. Leaching of Group IA and IIA ions leaves more highly charged ions in the soil. The pH decreases because the more highly charged metal ions like Ca^{2+}, Mg^{2+}, aluminum (Al^{3+}) and the transition metals are strongly attracted to water molecules. The attraction between the Al^{3+} and the oxygen in water weakens the O-H bond. The O-H bond is weakened so much that it sometimes breaks to release H^+ ions into the soil. This is illustrated below. The result is a more acid soil, and the pH goes down.

13. a. Nutrients are substances needed by green plants for healthy growth. There are 18 known elemental nutrients.
    b. Nonmineral nutrients are carbon, oxygen and hydrogen. They are available from water and atmospheric carbon dioxide.

c. Mineral elemental nutrients are water soluble substances that plants can only absorb as solutes through their roots.

Nonmineral	Primary	Secondary	Micronutrients
Carbon	Nitrogen	Calcium	Boron
Hydrogen	Phosphorus	Magnesium	Chlorine
Oxygen	Potassium	Sulfur	Copper
			Iron
			Manganese
			Molybdenum
			Vanadium
			Zinc

15. A soil shortage of N, P and K is more likely than a soil shortage of Ca, Mg and S. The nutrients N, P, and K are primary elemental nutrients and are needed in greater amounts. The elements N and K are typically present as very water soluble substances. They are easily leached from the soil.

17. The balanced equation for the reaction in the Haber process is:
$N_{2(g)} + 3 H_{2(g)} \rightleftharpoons 2 NH_{3(g)}$ This is a method to fix nitrogen from atmospheric nitrogen. It is the synthetic source of ammonia. The process is important because it provides cheap ammonia for agricultural fertilizer production on a huge scale. This supply of fixed nitrogen is essential for agricultural production in the world. World food production would not be able to meet the present population's needs if the Haber process was not available.

19. Chlorosis is a plant condition of low chlorophyll. It is caused by a deficiency of any one of the three nutrients: magnesium, nitrogen or iron. This deficiency leads to low chlorophyll content. A symptom of chlorosis is the presence of leaves that are pale yellow instead of green.

21. The fertilizer labeling system indicates the percentage by weight of nitrogen (N), phosphorus (P_2O_5) (phosphate) and potassium as K_2O (potash). None of the pure elements are present in the fertilizer. Each element is present in some compound. None of the reference substances are present in the fertilizer bag either.

23. Urea, H_2NCONH_2, decomposes in the soil to form ammonia.

25. a. A quick-release fertilizer dissolves readily in water. It dissolves easily and can be picked up quickly by the plant roots. A problem with this is that the fertilizer can be washed or leached out of the soil quickly.
    b. A slow-release fertilizer is not very water soluble. It dissolves very slowly in water and it is available to the plant more slowly, so it is taken up slowly by the plant.

© 1999 Harcourt Brace & Company. All rights reserved.

27. The positive side of ammonium nitrate is that it is inexpensive and a good source of water soluble nitrogen. Farmers benefit because cheap fertilizers keep food production costs down. Consumers benefit because food prices are low. The dark side of ammonium nitrate is that it is an explosive. Improper handling can lead to industrial accidents. It can be used as an explosive by terrorists. This was done in 1995 when approximately a ton of a fuel oil, ammonium nitrate mix was detonated next to the Federal Building in Oklahoma City.

29. DDT is a nonpolar organic compound. It is fat soluble because "like dissolves like" and both fats and DDT are nonpolar. DDT poses problems because it is a carcinogen and does not degrade quickly. It is relatively unreactive and is stored in the fat tissue of animals. This accumulation problem is exaggerated when one organism high in the food chain eats many other ones lower in the chain. The DDT levels in the higher organisms are compounded to higher levels because they acquire all the DDT from the lower organisms they eat.

31. About 33 % to 40 % of food crop production is lost to pests each year worldwide. The monetary value of these losses is estimated at $20 billion per year.

33. A biodegradable pesticide is one that is quickly converted to harmless products by microorganisms and natural processes. Pyrethrins are an example of a class of biodegradable insecticides. An example of a pyrethrin is shown here.

35. a. Pesticides can increase crop yields and protect them when stored. They reduce the loss of crops to pests. Every bit of food saved is food that need not be replaced. Food supplies increase and food is more plentiful at lower costs if pesticides are used. Malnutrition and starvation are reduced. Diseases carried by pests can be controlled or even eliminated. Epidemics of some diseases can be prevented or stopped. Pesticides can cause problems of water and soil contamination if they are misused. Pesticide residues can contaminate crops and create long term poisoning problems. Resistant strains of pests develop and require higher levels of pesticides.
    b. Pesticides should be used early enough so that smaller amounts are required and only when absolutely necessary. Other control measures should be used when possible and pesticides should be used only when alternative methods do not exist.

37. Sustainable agriculture is a set of practices intended to improve profits, limit use of agrichemicals and increase the use of environmentally friendly farming procedures.

39. Integrated pest management involves the use of disease resistant plant varieties and biological controls such as predators or parasites to control pests.

41. a. Transgenic crops are genetically altered crops with specific genes inserted to produce plants with desirable properties.
    b. Examples of transgenic crops are the FlavrSavr tomato, a weevil resistant garden pea and Bollgard cotton seed.

43. Pheromones are insect sex attractants. They are used to lure insects into traps so insecticide spraying is not needed.

45. Yes, there is a conflict. Population growth will decrease the number of productive acres per person. The present 0.82 acre per person will decrease and the efficiency of farming methods will need to improve in order to avoid famine and food shortages. This problem will be compounded because population centers usually are in the middle of prime agricultural land. The growth of urban areas will remove such land from production.

47. The LD_{50} values indicate toxicity. The LD_{50} is the dose that is fatal to 50% of the population. The greater the LD_{50} value the less toxic the substance. Low LD_{50} values indicate that the substance is more toxic because low doses are fatal for 50% of the population. In this list you can see that dimethin has very low toxicity. Dimethin has a LD_{50} of 10,000 mg per kg of body weight. This means that a 120 pound person would have to consume 545454 mg or 545.g of material to have a 50% chance of taking a toxic dose. This is more than a pound of pesticide.

Agrichemical	LD_{50}	
DDT	100 mg/kg	most toxic
primicarb	150 mg/kg	
carbaryl	250 mg/kg	
malathion	1,000 mg/kg	
dimethrin	10,000 mg/kg	least toxic

Solutions for Problems

1. $$\frac{\text{acres of land under cultivation}}{\text{number of people}} = \frac{4.84 \text{ billion}}{5.9 \text{ billion}} = \frac{4.84 \times 10^9 \text{ acres}}{5.9 \times 10^9 \text{ people}} = 0.82 \text{ acres/person}$$

2. The answers to this question are 10 % N and 2 pounds of nitrogen in a 20 pound bag. The label 10-20-20 means the fertilizer is 10 %N, 20 %P as P_2O_5, and 20 %K as K_2O. The 20-pound bag is 10 % N by weight. The 10 % means there are 10 pounds of N for every 100 pounds of total weight. The pounds of N are therefore

   $$\frac{10 \text{ pounds of N}}{100 \text{ pounds total}} \times 20 \text{ pound bag} = 2 \text{ pounds of N}$$

3. A fertilizer with 20-10-5 label is a complete fertilizer containing all three mineral nutrients listed in the order: nitrogen, phosphorus, potassium. On a 50 pound bag the 20-10-5 label means that 20% of the weight comes from nitrogen as N; 10% comes from phosphorus as P_2O_5; and 5% comes from potassium as K_2O. Therefore, 20% of the 50 pounds comes from nitrogen:

   $$50 \text{ pounds fertilizer} \times \frac{20 \text{ pounds N}}{100 \text{ pounds fertilizer}} = 10 \text{ pounds N}$$

Speaking of Chemistry

Name _____

Feeding the World

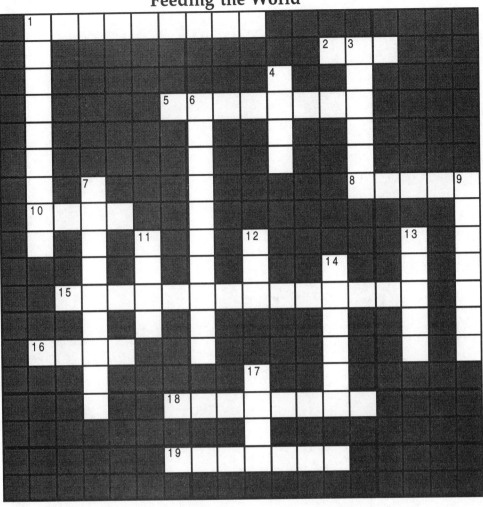

Across
1. Name for the anion PO_4^{3-}
2. Abbreviation for integrated pest management.
5. Fertilizer containing only one nutrient.
8. Decomposed organic matter in soil.
10. Fertilizer with formula, $(NH_2)_2CO$.
15. Nutrients required in very small amounts.
16. Mixture of mineral particles, organic matter, water and air.
18. Water _____ into the structure of particulate matter.
19. Nitrogen, phosphorus, and potassium are _____ nutrients.

Down
1. Natural pesticide.
3. K_2O
4. Calcium carbonate, $CaCO_3$.
6. Genetically altered crops.
7. Chemicals used to control pests.
9. Layer of soil that contains inorganic salts and clay particles.
11. Kind of pesticide that persists and breaks down slowly.
12. First chlorinated organic insecticide.
13. Condition with pH above 7.
14. Plants that can fix nitrogen.
17. Topsoil that is friable and has a high air content.

© 1999 Harcourt Brace & Company. All rights reserved.

Bridging the Gap   I   Name _____
Internet access to the U.S. Census Bureau and
Internet access to the Population Institute
Internet access to The Rockefeller University

This Bridging the Gap activity takes you into the Population Institute and the U.S. Census Bureau. You are supposed to look at the size of the current U. S. population and the size of the world population by checking the U. S. Census page. The Population Institute summarizes problems posed by continued global population growth, as does Rockefeller University

United States and World Population Today

To do this activity you are supposed to open the U.S. Census Bureau page with the following URL. This URL gives you two very simple population clocks. One clock is for the United States and the other is for the world. Use the clocks to get the estimated current populations. After you open this page you should be able to answer the following questions

        http://www.census.gov/main/www/popclock.html

        http://www.census.gov/main/www/popcld.html

1. What is the estimated population for the United States?

2. What is the estimated population for the world?

3. What percent of the earth's population lives in the United States?

© 1999 Harcourt Brace & Company. All rights reserved.

Population Institute

Use the following URL to access this site about population, soil, water and food supply information. After you open this page you should be able to answer the following questions.

http://www.populationinstitute.org/issue.html

4. How many people were added to the world population last year?

5. How many countries could not provide food for their population?

How Many People can the Earth Support?

This site points out that if present population growth rates continue into the future the world population would reach 694 billion by 2150. This raises the question: Can the world provide for this number of people?

http://www.rockefeller.edu/pubinfo/jecAAAS.nr.html

6. Based on the information presented on this page do you believe the population growth rates can be sustained until 2150? Give one factual piece of information to support your view.

Bridging the Gap   II   Name _____
Internet Link to The United States Geological Survey
Feeding the World

This Bridging the Gap takes you into the United States Geological Survey data on agricultural chemical use and farming practices in the U.S.

Agricultural practices in the United States

In this activity you are supposed to open the USGS page with the following URL.

http://h2o.usgs.gov/public/pubs/bat/bat000.html#HDR6

This URL gives you a very impressive page. The report consists of estimates for all counties in the conterminous United States of the annual use of 96 herbicides in 1989; annual sales of nitrogen fertilizer, in tons, for 1985-91; and land use, chemical use, and cropping practices in 1987. The information can be used to estimate regional agricultural chemical use across the United States. You should scroll to the following line entry and click.
Figure 1. Estimated annual county-level herbicide use, 1989: (a) atrazine; (b)alachlor.
This line has the following URL. This URL gives a map of the United States.

http://h2o.usgs.gov/public/pubs/bat/fig1.gif

The map shows the usage of atrazine in pounds per square mile for every county in the United States.

You are supposed to find your state, read the level of atrazine herbicide use and record the amount on the Concept Report Sheet. Examine the map showing atrazine use and determine which region of the 6 below used the greatest number of pounds of atrazine per square mile. Identify the region that used the least atrazine per square mile.

Southwest   Central Midwest   Northeast   Northwest   Northern Midwest   Southeast

Additionally, your assignment is to open the following URL to access another U.S. map. This one shows the usage of potash fertilizer in pounds per square mile for every state in the U.S.

http://h2o.usgs.gov/public/pubs/bat/fig6.gif

You are supposed to find your state or one nearby, read the level of potash use and record the value on the report sheet. Additionally determine which region of the country used the greatest number of pounds of potash per square mile. Identify the region that used the least potash per square mile.

© 1999 Harcourt Brace & Company. All rights reserved.

Bridging the Gap
Feeding the World
II   Name _____

Agricultural practices in the United States: Herbicides
What is the level of atrazine use in your state or one nearby? Record the amount here.

Which region of the country used the greatest number of pounds of atrazine per square mile.

Southwest   Central Midwest   Northeast   Northwest   Northern Midwest   Southeast

Identify the region that used the least atrazine per square mile.

Southwest   Central Midwest   Northeast   Northwest   Northern Midwest   Southeast

Which part of the country seems to have a greater problem controlling weeds? What do you think causes this?

Agricultural practices in the United States: Fertilizers
What is the level of potash (K_2O) use in your state or one nearby? Record the amount here.

Which region of the country used the greatest number of pounds of potash per square mile.

Southwest   Central Midwest   Northeast   Northwest   Northern Midwest   Southeast

Identify the region that used the least potash per square mile.

Southwest   Central Midwest   Northeast   Northwest   Northern Midwest   Southeast

Which part of the country seems to have a greater problem maintaining potassium in soil? What do you think causes this? Do you think that leaching has any bearing on this?

© 1999 Harcourt Brace & Company. All rights reserved.

Speaking of Chemistry Crossword Puzzle Solutions

Chapter 1

Living in a World of Chemistry

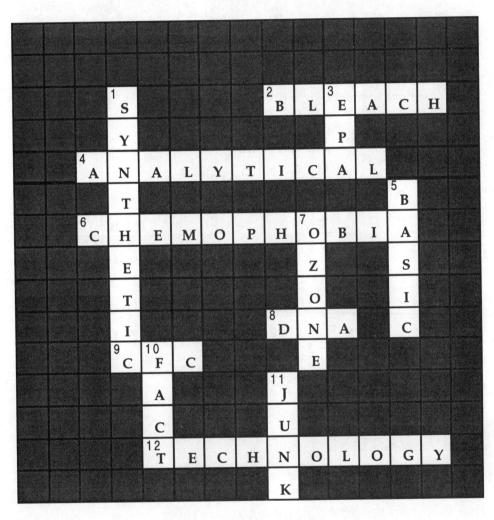

Chapter 2

The Chemical View of Matter

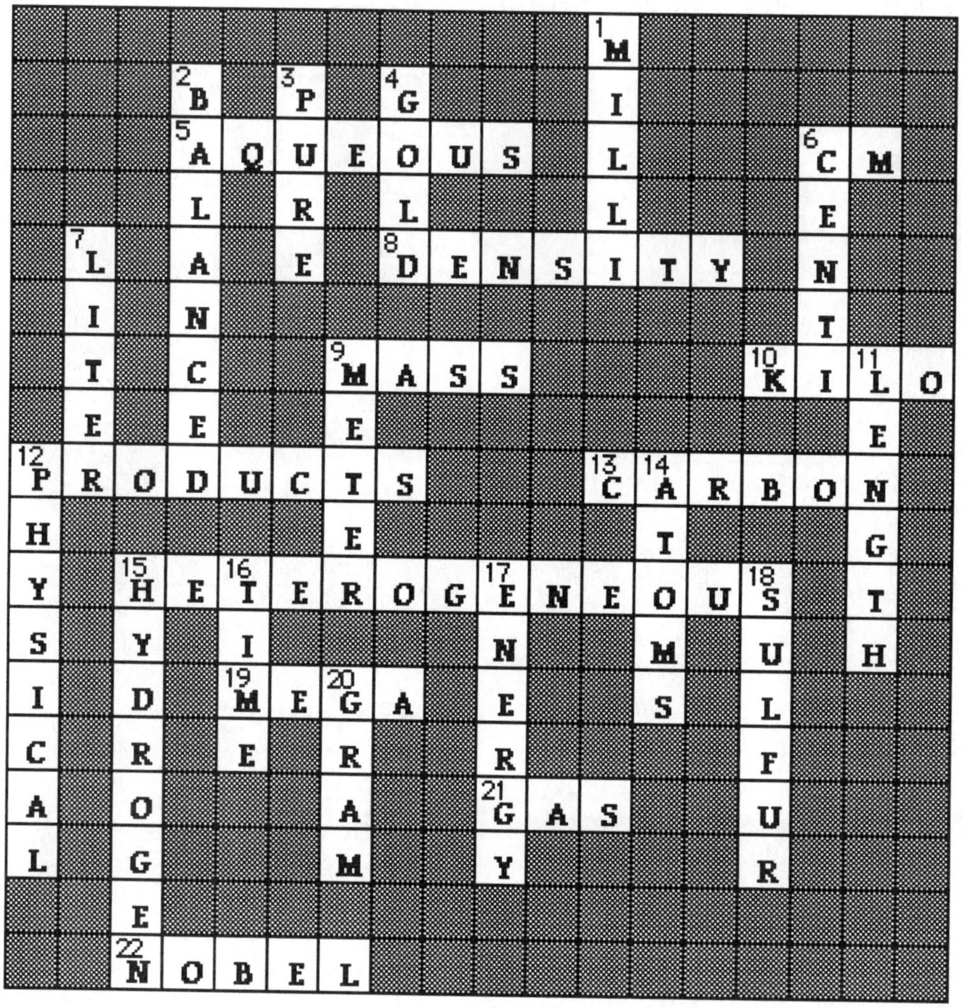

1 M
2 B 3 P 4 G I
5 A Q U E O U S L 6 C M
L R L L E
7 L A E 8 D E N S I T Y N
I N T
T C 9 M A S S 10 K I 11 L O
E E E E
12 P R O D U C T S 13 C 14 A R B O N
H E T G
Y 15 H E 16 T E R O G 17 E N E O U 18 S T
S Y I N M U H
I D 19 M E 20 G A E S L
C R E R R F
A O A 21 G A S U
L G M Y R
E
22 N O B E L

Chapter 3

Structure of the Atom

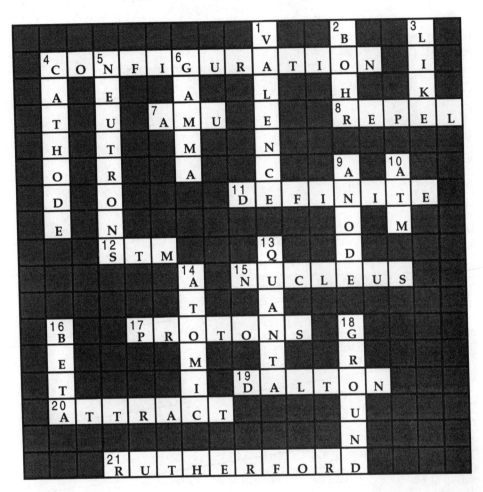

Chapter 4

Nuclear Changes

							¹B	²E	³T	A		⁴R	
				⁵C				I		⁶L	E	A	D
		⁷P		U		⁸H	A	H	N		P		D
		L		R				S			H		O
⁹U	R	A	N	I	U	M		T			A		N
		S		E				E					
		M				¹⁰P		I		¹¹F			
	¹²B	A	C	K	G	R	O	U	¹³N	D			
	I					S			¹⁴E	P	A		
¹⁵C	A	N	C	E	R			I		C			
	D					T		¹⁶J	A	P	A	N	
¹⁷M	I	N	U	S		R		Y					
	N			¹⁸Z		O							
	G		¹⁹C	H	E	R	N	O	B	Y	L		
				R									
		²⁰I	S	O	T	O	P	E					

Chapter 5
Chemical Bonding & States

Chapter 6

Chemical Reactivity

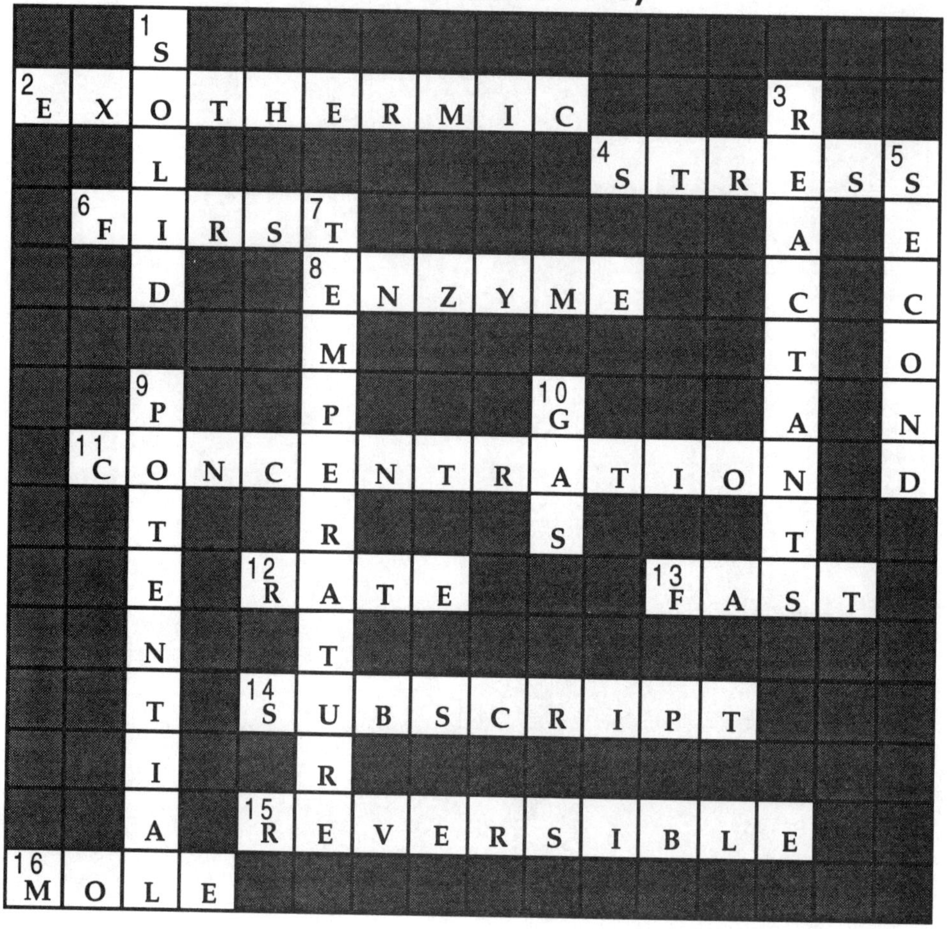

Across:
2. EXOTHERMIC
4. STRESS
6. FIRST
8. ENZYME
11. CONCENTRATION
12. RATE
13. FAST
14. SUBSCRIPT
15. REVERSIBLE
16. MOLE

Down:
1. SOLID
3. REACTANT
5. SECOND
7. TEMPERATURE
9. POTENTIAL
10. GAS
13. F

Chapter 7

Acid–Base Reactions

	¹P		²C	O	L	³O	R							
	R		A			N				⁴S		⁵B		
	O		U			E		⁶H	⁷Y	D	R	O	N	I	U	M
	T		S	⁸W				Y	I		U		F	
	O		T	E				D	E		R		F	
⁹I	N	D	I	C	A	T	O	R				¹⁰R	E	D
			C	K		¹¹O	N	E				R
				¹²S		X				¹³M		
	¹⁴N	¹⁵E	U	T	¹⁶R	A	L	I	¹⁷Z	A	T	I	O	N
		L			R		C		D		N		L	
		E			O		I		E		T	¹⁸B	A	S	E
¹⁹S	E	V	E	N			D				A	R		
		E			G						C		²⁰F	
		N					²¹K				I	²²T	W	O
				²³B	L	O	O	D				Y		U
							H							R

279

Chapter 8

Oxidation–Reduction

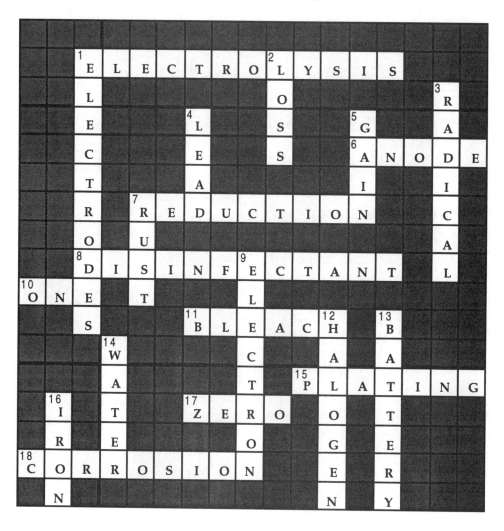

Chapter 9

								¹P		
	²M			³F	O	U	R			
⁴O	C	T	A	N	E			⁵O	I	L
R		B			I			P		
T		⁶E	T	H	A	N	E	A		
H								N		
⁷O	X	Y	⁸G	E	N			E	⁹C	I	¹⁰S
			T		¹¹C					I
	¹²M		¹³H	Y	D	R	O	¹⁴C	A	R	B	O	N
	E		A					L		G
	¹⁵T	R	A	N	S		L		L
	H		O		¹⁶O		¹⁷P	A	R	A	E
	A			¹⁸A	L	K	E	N	E	
	N				E		¹⁹M	E	T	A
	²⁰E	T	H	E	R					

© 1999 Harcourt Brace & Company. All rights reserved.

Chapter 10

				¹B			²H	³D	P	E			
⁴P	⁵R	I	M	A	R	Y		V				⁶R	
	U			N			⁷C	⁸L	P	E		E	
	B			A		⁹R		D				C	
	B		¹⁰N	Y	L	O	N					Y	
	E		A			¹¹X		A		P		C	
	R					Y		D		E		L	
		¹²P				G		¹³V		C	¹⁴C	I	
¹⁵S	T	E	R	E	O	¹⁶E	G	U	L	A	T	I	O	N
T		T				N		L		L		S		G
Y				H				C						
R				E			¹⁷A	L	D	E	H	Y	D	E
E			¹⁸P	R	O	O	F							
N							¹⁹I	S	O	P	R	E	N	E
²⁰E	T	H	Y	L	E	N	E							
							Z							
			²¹F	O	R	M	A	L	D	E	H	Y	D	E

282

© 1999 Harcourt Brace & Company. All rights reserved.

Chapter 11

The Chemistry of Life

					¹P	²A	S	³T	E	U	R				⁴S
						O		A					⁵R	N	A
						A		⁶A	U	G					C
					⁷A	L	P	H	A						C
					T										H
	⁸N	⁹N	O	N	P	O	L	A	R			¹⁰P	O	L	A	R
		P										Y			I
		T			¹²U		¹³P				¹⁵S			N
¹⁴D	I	S	A	C	C	H	A	R	I	D	E	S			
	C			U			O		N		I			
¹⁶F	A	T	S	¹⁷C		T		A		N				
L				O		E				E				
	¹⁸L			D		I								
¹⁹H	Y	D	R	O	G	E	N	²⁰A	T	E	D			
E				N				D						
			²¹L	I	P	I	D	S						

Chapter 12

Nutrition: The Basis of Healthy Living

	¹C	H	O	L	E	²S	T	E	R	O	L				
						C							³A		
						U		⁴R			⁵F		N		
			⁶C			R		⁷A	M	Y	L	A	S	E	
		⁸S	A	L	I	V	A		D		T		⁹M	S	¹⁰G
			R			Y		¹¹F	I	B	E	R		I	L
	¹²P		B						C			¹³F	D	A	U
	L		O		¹⁴G				A						C
	¹⁵A	T	H	E	R	O	¹⁶S	C	L	E	R	O	S	¹⁷I	S	O
	Q		Y		A		U							O		S
	U		D		S		G		¹⁸M	E	G	A	D	O	S	E
	E		R				¹⁹A	T	P					I		
			A				R				²⁰R	A	N	C	²¹I	D
			²²T	E	N		²³S	A	F	E				E		R	
			E												O
						²⁴A	N	T	I	O	X	I	D	A	N	T

Chapter 13

Chemistry and Medicine

			¹F			²A					
	³P	⁴D	O	P	⁵A	M	I	N	E		
	A	A		B	D			⁶C			
	T	A		U	S		⁷O	P	O	I	D
	H			S				C			
	O	⁸F	E	V	E	R		A			
	G	O			D		⁹A	N	G	I	N	A
	E	L						N			
¹⁰A	N	T	I	M	E	¹¹T	¹²A	B	O	L	I	T	E	S
Z		C				S	T				
¹³T	¹⁴P	A					¹⁵C	O	D	E	I	N	E
	L		¹⁶P				I				
¹⁷G	A	B	A				R				
	S		I		¹⁸S		I		¹⁹P		
	M		N	²⁰P	E	N	N	I	C	I	²¹L	L	I	N	G
	I			F					P		S	
²²C	A	N	C	E	R						D	

Chapter 14

							¹M					
					²C	A	S	T		³Z		
			⁴T		G					E		
⁵S	U	P	E	R	C	O	N	D	U	C	T	O	R
E			A				E			R		
A			N			⁶H			O		
	⁷G	⁸L	A	S	S		⁹I	G	N	E	O	U	S		¹⁰S		¹¹P
		I					U			L			T		Y
		M				¹²M	R	I		E			E		R
		E								S			E		O
		S					¹³P					L		X
		T		¹⁴R	O	¹⁵A	S	T	I	N	G				E
		O				L				G				N
	¹⁷D	U	C	T	I	L	E		¹⁸R	E	D	U	C	E	D
					O				E					S
				¹⁹B	R	I	N	E	S
				Y				
				S				

286

Chapter 15

Water--Plenty of It, but of What Quality?

	1	2	3	4	5	6	7	8	9	10	11
1	B			C	A	R	C	I	N	O	G	E	N

Across / Down entries:

- 1D: BROMATE
- 2A: CARCINOGEN
- 3D: ACC... / ACID...
- 4D: EPA
- 5A: OSMOSIS
- 6D: POTABLE
- 7D: BOD
- 8D: SULFHYDRYL
- 9A: DILUTION
- 10A: CARBONBLACK
- 11D: ARSENIC
- 12D: CALCIUM
- 13A: LEAD
- 14A: SETTLING
- 15D: RCA
- 16D: HRD
- 17A: RECHARGE
- 18A: DDT

Chapter 16

Air: The Precious Canopy

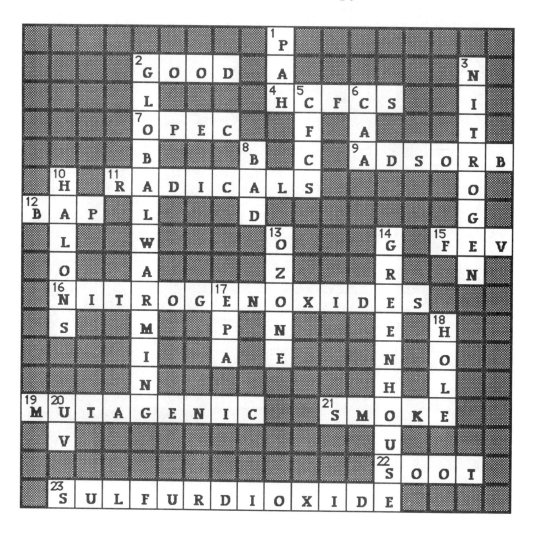

Chapter 17

Feeding the World

Across/Down	Answer
1 Across	PHOSPHATE
1 Down	PYRETHRUM
2 Down	IPM
3 Down	POTAS...
4 Down	LEME...
5 Across	STRAIGHT
6 Down	TRANSGED
7 Down	PSTA...
8 Across	HUMUS
9 Down	SUBSOIL
10 Across	UREA
10 Down	UM
11 Down	HA
12 Down	DD
13 Down	BASIC
14 Down	LEGUM
15 Across	MICRONUTRIENTS
15 Down	MCDE
16 Across	SOIL
17 Down	LA
18 Across	ABSORBED
19 Across	PRIMARY

© 1999 Harcourt Brace & Company. All rights reserved.